The Moon

Its Creation,

Form

and

Significance

The Moon

Its Creation,

Form and Significance

John C. Whitcomb, Th.D.

Donald B. DeYoung, Ph.D.

BMH Books

Winona Lake, Indiana 46590

ABOUT THE AUTHORS—

JOHN C. WHITCOMB has been actively involved in the study of origins as related to the Book of Genesis since 1942. Two early books published were: *The Genesis Flood: The Biblical Record and Its Scientific Implications*, with Henry M. Morris; and *The Origin of the Solar System*. He has also written *The Early Earth* and *The World That Perished*, created several cassette courses, and published four Bible charts. Dr. Whitcomb is professor of theology and Old Testament, and director of postgraduate studies at Grace Theological Seminary, Winona Lake, Indiana. He is a graduate of Princeton University and Grace Theological Seminary.

DONALD B. DeYOUNG is a physicist who is interested in the areas of solid-state and nuclear science, as well as astronomy. Several of his articles have been published in *The Journal of Chemistry and Physics of Solids*, *The Journal of Chemical Physics*, and the *Creation Research Society Quarterly*. Dr. DeYoung is associate professor of physics at Grace College, Winona Lake, Indiana, and is a graduate of Grand Rapids Junior College, Michigan Technological University, and Iowa State University.

Cover photo: The Apollo 11 lunar module ascent stage during rendezvous with the command service module. The earth rises above the lunar horizon. (NASA photo, cover design by Timothy Kennedy.)

ISBN: 0-88469-102-0

Dedication

To Norma, my wife,
and our children,
    Dan
    Dave
    Tim
    Don
    Connie
    Bob

John C. Whitcomb

To Sally, my wife,
and our children,
    Jenny
    Marjorie
    Jessica

Donald B. DeYoung

Table of Contents

FOREWORD 9

INTRODUCTION 13

PREFACE 21

ACKNOWLEDGMENTS 25

CHAPTER I.   LUNAR EXPLORATION 29

CHAPTER II.   NATURALISTIC THEORIES OF THE
              MOON'S ORIGIN 35

The Fission or Break-Away Theory 37

The Capture Theory 42

The Nebular or Condensation Theory 45

CHAPTER III. THE GENESIS RECORD OF THE MOON'S
              CREATION 53

The Double-Revelation Theory 54
    Limitations of the Scientific Method 55
    Scientific Obstacles to Cosmic Evolutionism 56
    A Fatal Neglect of Biblical Revelation 62

The Creation of the Moon 69
    The Manifestation of God's Wisdom and Power
        Through the Direct Creation of the Moon 70
    The Instantaneous Creation of the Moon 73

CHAPTER IV.   LUNAR GEOLOGY 85

Lunar Rocks 86
    Crystalline Rock 87
    Lunar Soil 88
    Breccia 88
    Lunar Rock Analysis 89

Hypothetical Moon History 91

Radiometric Dating Evaluation 98

CHAPTER V.   TRANSIENT LUNAR PHENOMENA   105
  Historical Background   105
  The Moon's Dark Side   110
  The Mountains   112
  The Craters   114
    The Crater Aristarchus   114
    The Crater Plato   116
    The Crater Alphonsus   117
  The Maria   119
    Mare Crisium   119
    Mare Vaporum and Mare Nubium   120
  Some Proposed Mechanisms   121
CHAPTER VI. LUNAR DISTINCTIVES   129
  Introduction   129
  Lunar Brightness   130
  Solar Eclipses   132
  The Moon's Shape   136
  Ocean Tides   138
  Further Lunar Properties   140
    Orbital Angular Momentum   141
    Lunar Phases   141
    Lunar Inclination   141
    Orbital Synchronism   145
APPENDIX I.   THE DEPENDABILITY AND DESTINY
              OF THE MOON   147
APPENDIX II.   MOON WORSHIP—A SPIRITUAL
               DISASTER   153
  The Theological Significance of the Moon's
  Creation on the Fourth Day   153
  The Old Testament and Moon Worship   155
  The Worship of the Moon in the Ancient Near East   157
APPENDIX III.   THE BIBLE AND SCIENCE—A Spectrum
                of Representative Writings Among Christian
                Theologians and Scientists   163
  Representatives of the Double-Revelation Perspective   163
    Scientists   163
    Theologians and Philosophers   165
  Representatives of the Recent-Creation Perspective   166
    Scientists   166
    Theologians and Philosophers   169
INDEX OF SUBJECTS   171
INDEX OF NAMES   175
INDEX OF SCRIPTURE   179

LIST OF TABLES AND FIGURES

Tables

I-1   Summary of Major Lunar Exploration Events   32

II-1   Summary of Major Problems of Naturalistic Lunar Origin Theories   49

III-1   The Earth's Magnetic Dipole Moment Plotted against Time.   60

IV-1   A Classification Diagram of Returned Lunar Samples . .   87

IV-2   Three New Igneous Rock Minerals Found on Mare Tranquillitatis   89

IV-3   Major Postulated Geological Events on the Moon   92

IV-4   Variation in Ages for Apollo Sample Material   100

VI-1   Physical Properties of the Earth and Moon   143

VI-2   A Comparison of Natural Satellite Masses   144

VI-3   An Angular Momentum Comparison for Planets with Satellites   144

Figures

II-1   The Fission ("daughter") Theory   37

II-2   The Capture ("spouse") Theory   43

II-3   The Ellipse of Satellite Orbits   45

II-4   The Nebular ("sister") Theory   46

V-1   The Moon's Near Side Divided into Quadrants   109

V-2   Major Features of the Moon at Full Phase   115

VI-1   Sizes of the Larger Satellites Compared to Their Planets   131

VI-2   The Sun and Moon Appear to Be the Same Size in the Sky   132

VI-3   The Size of the Moon Compared to the United States . .   137

VI-4   Earth Tides Caused by the Moon   139

VI-5   The Phases of the Moon   142

A II-1   A Reconstruction of a Babylonian Ziggurat   159

A II-2   A Pillar with Hands Raised toward Moon   161

Foreword

The Apollo moon project spawned popular interest and intensive research efforts of a nature heretofore unknown in astronomy. Never before had astro-geophysicists been accorded the possibility of engaging in real specimen, laboratory studies or of using on-site instrumentation to explore an astronomical body in a manner remotely comparable to that known in terrestrial geophysics.

This book, in a similar way, provides a profoundly different perspective to lunar studies in particular and astronomy in general. Current scientific writings commonly address the topic by dealing exclusively with naturalistic speculations for the existence of the moon together with a discussion of the present state of the lunar surface and events observed thereon. The authors here go a step beyond these familiar approaches and discuss the origin, state, and observables of the moon simultaneous with its purpose, significance, and eventual destiny. Such a combination of topics appearing in a single context is unique in astronomical literature and adds an illuminating dimension to any study pertaining to origins.

By the title alone, the authors manifestly declare their underlying tenet which is carefully and distinctly elaborated in the text; i.e., they accept the entire Biblical record as authoritative in regard to beginnings, history, science, and ultimate meanings. The Bible teaches explicitly that the moon was created instantaneously as a functioning body in the heavens and at a time simultaneous with the sun, planets, stars, and galaxies, but three days subsequent to the creation of the earth. Furthermore, the Bible states clearly the intended purpose for the moon's existence and its proximity to the earth. Their literal acceptance of these truths is amply evident in the text and comprises the cornerstone on which the geology, lunar data, observable phenomena, and origin of the moon are discussed.

Prevailing scientific theories for the moon's existence all begin with the presupposition that this enigmatic body, and the earth also, are four or five billion years old. These theories are naturalistic; that is, they depend entirely on the application of natural law as discerned to be operating at present and are devoid of any external or supernatural action. In contradistinction to popular explanations, the authors here argue for a relatively recent origin for the moon by special creation through the action of divine omnipotence. Although science compels neither view, the present work documents the failure, or at best the inconclusiveness, of competing theories and shows that the paucity of firm evidence supporting any one theory conforms more naturally to the creationist view. All readers may not agree with this assessment nor with the scientific and/or Biblical interpretations expounded by the authors. Nevertheless, their arguments are carefully reasoned and are supported by an abundance of significant and recent documentary evidence. Both the scientific and Biblical discussions are sufficiently comprehensive and objective to warrant an honest hearing in the classroom, among truth-seeking scientists, and also in the church.

Science divorced from Biblical authority can provide some details about the present state of the moon (or any other observable), but, by its very nature, science alone can never be authoritative when questions dealing with the "why" and "how" of origins are encountered. More significantly, the purpose and meaning of existence are outside the realm of science. Such questions are commonly thought to belong in more abstract disciplines or metaphysics. However, it is my opinion that the basic premise and approach of the authors remove a great deal of the subjectivism present in such disciplines and provide a framework within which seemingly disparate topics such as origin and purpose, factual evidence and true significance, etc., can be productively and effectively studied. These important topics were not intended by the Creator to be investigated in isolation and the authors, by combining sound Biblical and scientific scholarship, have produced a treatise of distinguished quality and a valuable perspective from which future studies of our fascinating universe can be launched.

Larry G. Redekopp, Ph.D.
Associate Professor of Aerospace Engineering
University of Southern California
Los Angeles, California 90007

Introduction

Throughout recorded history men have been fascinated by the presence and significance of the moon. But never has such interest been so intense as in our present day, with pieces of the moon now on earth and human footprints and machinery on the moon. Possibly two billion people have watched or read or heard about the lunar landings. But no segment of human society should be more interested in the meaning of these events than the Christian community which uniquely possesses the only Book that can give ultimate answers to questions concerning cosmic origins.

The Bible has much to say about the moon and the universe beyond it. Therefore, one might think that the ideal time for publishing a serious study of the moon from a Biblical and scientific perspective would have been shortly after the completion of the great Apollo mission. In retrospect, however, the world was so inundated with masses of raw data concerning the moon in the early 1970s that a psychological reaction inevitably set in.

> That feat [a human foot first touching the moon], the President of the U.S. assured his countrymen, was to be ranked as the greatest thing since—*Creation.* After that exaltation, there was only one way, by the law of psychological gravity, for Sci-Tech's prestige to go. Sure enough, down it went. And in its place has arisen a new public attitude that seems the antithesis of the former awe.[1]

This is surely an overstatement; but it is safe to say that

---

[1]Frank Trippett, "Science: No Longer A Sacred Cow," *Time* 109:10 (March 7, 1977), 72.

now, a decade after the first Apollo landing, the time for a careful reappraisal has surely come. It has required several years for lunar scientists even to begin to dig into the incredible mass of newly acquired data and to offer some tentative conclusions. Unfortunately, however, there can be no "whole picture" of our moon until it is also seen within the framework of the infallible Word of the Creator God.

Such a statement will sound strangely medieval to many ears. What significant light can the Jewish and Christian Scriptures shed on such questions as the origin, significance, and destiny of the moon? Are not these the kind of questions that must be answered exclusively by those who have examined the empirical data? How could a book written more than two thousand years ago, long before the Copernican revolution and the invention of the telescope, be worthy of serious consideration by modern astronomers who have dedicated themselves to the search for solid answers to these haunting questions?

It is the foundational premise of the present volume that Christian theology (exegetically based upon the Hebrew and Greek texts of the Bible) is absolutely crucial for meaningful answers as mankind searches the universe that surrounds him. The Bible has indeed been quoted on occasion by leading astronomers. But why? Usually, perhaps, for aesthetic or emotional effect. At best, for devotional insights concerning the wonder of it all. But the latter constitutes only a very small part of the reason why the Bible was given to us by God. The primary purpose of the Bible is *doctrine* (Gk. *didaskalia*, "teaching," with emphasis on objective data concerning God and His dealings with men). In the key statement of the New Testament on the nature of the Bible, doctrine is the very first reason why this revelation has been given: "All scripture is given by inspiration of God, and is profitable for doctrine . . . ." (2 Tim. 3:16). This essential priority has been largely lost upon a generation that has been strongly influenced by neo-orthodox existentialism. It was not for emotional, aesthetic, or even devotional reasons that *our Lord Jesus Christ referred to each of the first seven chapters of Genesis and every New Testament writer referred to Genesis 1-11.* Christ and the apostles not only took for granted, but insisted on the absolute historicity and authority of those opening chapters of the Bible.

For all practical purposes, however, the vastly important science of astronomy has been all but abandoned to evolutionary interpretations by most Christians for well over a century. R. Hooykaas, professor of the History of Science at the University of Utrecht, points out that during the nineteenth century "theological opposition to cosmogonic hypotheses as to the origin of the moon and other heavenly bodies, remarkably enough, never was very great, even when the 'natural' explanation of their birth required long periods."[2]

Such indifference to the Biblical doctrine of creation is all the more astounding in our own day in view of the repeated and widely publicized failure of astronomers to present even remotely possible mechanisms for the evolution of our nearest astronomical neighbor, the moon. Could the Creator be speaking to the conscience of His Church through such scientific frustrations?

The authors desire at the very outset to make their positions clear concerning the relation of Scripture to natural science. It is their profound conviction that there can be no *ultimate* contradiction between these two media of divine communication to the minds of men. Apparent contradictions do exist. In the very nature of the case this is inevitable, and serves as an eloquent and humbling testimony to the limitations of the human mind even in the "space age." No contradictions can be resolved in the mind of true Christians by relegating them to the "pre-scientific world view" of Biblical writers, for they were simply the spokesmen of the divine Author who inspired them (cf. Matt. 5:18; John 10:35; 2 Tim. 3:16; 2 Peter 1:21). A century of spectacular archaeological vindications of historical references in the Bible should be sufficient to give pause to those who would claim otherwise.[3]

---

[2] R. Hooykaas, *The Principle of Uniformity in Geology, Biology and Theology* (Leiden: E. J. Brill, 1963), p. 205.

[3] For example, references to Belshazzar and Darius the Mede in the Book of Daniel had been challenged for centuries by critics of the Bible. Archaeological

On the other hand, scientific investigation plays a vital role in the plan and purpose of God for mankind. The first human beings were commanded by God to "subdue" the earth and to exercise "dominion" over it (Gen. 1:28). Solomon observed that "it is the glory of God to conceal a thing: but the honour of kings is to search out a matter" (Prov. 25:2). God, being omniscient, can discover nothing new. So it has pleased Him to create a universe filled with things that are mysterious and challenging to the minds of men created in His image and likeness. And since those who "search out a matter" are described as "kings," we may say that science is a royal activity in the sight of God.

Some have insisted that Biblical Christianity has been a hindrance to the advancement of science.[4] But wiser men have come to an exactly opposite conclusion.[5] Some of the greatest pioneers in the history of modern science have been devoted Christians.[6] The contemporary tensions between Biblical creationists and natural scientists have come about, therefore, through a basic misunderstanding. It has been widely assumed that while Biblical creationism involves a subjective faith-commitment to a world and life view (cf. Heb. 11:3), the consensus of modern science on origins is the end product of purely unbiased objectivity. The utter naiveté and, in fact, absurdity of such a contrast is now, at long last,

---

discovery has brought completely new light to bear upon this supposedly hopeless contradiction. Cf. J. C. Whitcomb, *Darius the Mede: The Historical Chronology of Daniel* (Nutley, N.J.: Presbyterian and Reformed Publishing Co., 1963).

[4]For a rather notorious example, see Andrew D. White, *A History of the Warfare of Science With Theology in Christendom* (New York: George Braziller, reprinted 1955). White, a noted theological liberal, was the first president of Cornell University (1867-85). The first edition of this book appeared in 1896.

[5]"Contrary perhaps to what one would have expected, a more fully Biblical world view has, since the sixteenth century, favoured the rise of modern science and of the world picture connected with it" (R. Hooykaas, *Religion and the Rise of Modern Science* [Grand Rapids: Wm. B. Eerdmans Publishing Co., 1972], p. 13). See also Robert E. D. Clark, *The Christian Stake in Science* (Chicago: Moody Press, 1967).

[6]Among these have been Johannes Kepler (1571-1630), Robert Boyle (1627-91), Michael Faraday (1791-1867), Samuel F. B. Morse (1791-1872), Lord Kelvin (1824-1907), and James Clerk Maxwell (1831-79). Cf. Robert E. D. Clark, *Christian Belief and Science* (Philadelphia: Muhlenberg Press, 1960), pp. 64-74.

becoming clear to a large number of philosophers.[7] By the very nature of the case, a concept of ultimate origins involves basically unprovable presuppositions that are religious in character. Honesty and wisdom require that these presuppositions be recognized and stated at the outset. Christians must do this because the Bible requires them to do so in the name of their God. But proponents of evolutionary cosmogonies have generally fallen short of the basic requirements of objectivity and honesty in this regard.

Science, then, can be a magnificent and royal enterprise in proportion to its acknowledgment of the supremacy of the King. That is exactly the tragedy of the modern impasse. Rejecting the claims and rights of the great King of Creation, modern minds, in the name of human autonomy, have not only rejected His Word, but have ignored the subjectivism that has permeated that denial. Turning from their King, they have used the very science that He alone has made possible to ignore His sovereignty in the universe! In the words of Isaiah: "You have forgotten the Lord your Maker, who stretched out the heavens, and laid the foundations of the earth" (Isa. 51:13). The issues are, at bottom, moral and ethical, rather than merely academic and intellectual.

In keeping with these stated presuppositions, the authors have sought, in an admittedly tentative and preliminary fashion, to structure lunar science within the framework of Biblical revelation. An opening chapter dealing with the contemporary setting of lunar discoveries is followed by two chapters that deal with the vastly important question of origins. The rather disastrous failure of secular speculations serves as a significant backdrop for the unchanging simplicity and grandeur of the Biblical record of the moon's creation. Vital to this entire discussion is the "double-revelation" theory, which has all but eliminated the clear message of the Genesis creation account in the thinking of many Christian philosophers and scientists in recent years.[8]

Chapters IV and V are devoted to an analysis of the actual structure, composition, and functions of the moon. The im-

---

[7] For a representative list, see Chapter III, note 4.

[8] See Appendix III for a partial list of those who have advocated this theory.

plications of these facts for the comparatively recent creation
of the moon are presented, especially with regard to the lack
of thick moon dust layers, the presence of a molten interior,
and the dating of lunar soil. Chapter VI shifts the emphasis
from recent origin to purposeful design. The size, shape, distance, and albedo of the moon all contribute to the well-
being of mankind in numerous and remarkable ways. Most
people find it quite impossible to believe this can be explained by pure chance. But there are exceptions. In a
volume entitled, *The Atheism of Astronomy: A Refutation
of the Theory That the Universe Is Governed by Intelligence*
(one of a series of books edited by Madalyn Murray O'Hair
under the general title, *The Atheist Viewpoint*), Woolsey
Teller flings out the challenge:

> The teleologist who sees "design" where there is none should
> explain what "purpose" is served by the moon other than to raise
> huge tidal waves on the earth and drown people in floods. The
> moon is a mere "drag" on the earth. It does not support life
> itself, and its senseless revolutions around us are quite in keeping
> with the whole futility of cosmic movement.[9]

The authors humbly suggest, in light of the evidences presented in Chapter VI, that this pre-Apollo assertion is far less
credible now than it was when first published.

Three appendices conclude the volume. The first surveys
the Biblical revelation concerning the moon's basic functions
as a calendar and especially as a sign of the faithfulness and
sovereignty of God. The rare supernatural interruptions of its
normal functions (especially in connection with eschatological events) only serve to accentuate the personal concern of
the Creator for the inhabitants of the earth in terms of mercy
and judgment.

The second appendix focuses attention on the spiritual
disaster of moon-worship. This clearly and frequently stated
divine warning places the entire debate concerning the order
of events during Creation Week into a new perspective. The
apparent absurdity of an initial creation of the earth followed
by the creation of the sun, moon and stars gives way to a

[9]Woolsey Teller, *The Atheism of Astronomy* (New York: The Truth Seeker
Company, 1938), p. 94.

deeper purpose, typically divine, of accentuating the infinite distinction between the Creator and the creation, including the astronomical creation. Thus, each Biblical teaching is seen to supplement and illuminate other Biblical teachings; and to deny one teaching of the Bible in order to make peace with a contemporary scientific consensus is to distort its entire fabric. In the end, the authors firmly believe, the scientific consensus will be in total harmony with every detail of Biblical revelation. Granted, this is a faith projection. It cannot be absolutely demonstrated to the complete satisfaction of every mind. But it is a faith that has been remarkably confirmed by the discoveries of the Apollo mission. And, by contrast, the various evolutionary concepts, which are clearly contradicted by the Genesis account, have been rendered more incredible than ever before by what we now know about the moon.

The third appendix is intended to provide a bibliographic foundation and stimulus for further research into these crucial questions. If the authors have succeeded, to some extent, in awaking interest in the Biblical concept of origins, their labors will be fully justified. Constructive criticisms by qualified readers will be welcomed, for the present work is only a small beginning. But if God and His Word have been truly honored, His ultimate blessing may be confidently anticipated.

Preface

During the late 1960s and early 1970s the modern creationist movement (as it has come to be known) developed rapidly in breadth of outreach and depth of quality. Outstanding studies were being produced in the areas of geology, paleontology, hydrodynamics, biochemistry, genetics, thermodynamics and educational philosophy from a strictly creationist perspective. In addition, remarkable success was being experienced in university debates in exposure through mass media, and even in legal and political areas. Judging from exhibitions of emotional over-reaction, the evolutionary establishment was being deeply shaken.[1]

However, during this period of remarkable growth for scientific creationism, one major area of continued neglect was astronomy. Several technical articles appeared from time to time.[2] But the need was being increasingly felt for a definitive work on astronomy from the perspective not only of science but also of Biblical theology and exegesis.

In the fall of 1974 it became evident to the present authors that God's time had come for such a project to be

---

[1]See, e.g., Norman D. Newell, "Evolution Under Attack," *Natural History* 83:2 (April, 1974), 32-39; and Preston Cloud, " 'Scientific Creationism'—A New Inquisition Brewing?" *The Humanist* 37:1 (January/February, 1977), 6-15. For a reply to Preston Cloud's article, see *Acts and Facts* [Institute for Creation Research, 2716 Madison Ave., San Diego, CA 92116] 6:3 (March, 1977), pp. 1-2, 7.

[2]Cf. George Mulfinger, Jr., "Review of Creationist Astronomy," *Creation Research Society Quarterly* [hereinafter referred to as *CRSQ*] 10:3 (December, 1973), 170-75. See also Gerardus D. Bouw, "The Rotation-Curve of the Virgo Cluster of Galaxies," *CRSQ* 14:1 (June, 1977), 17-24.

launched. Each is profoundly convinced that God has providentially prepared him through the years for such a study as this. One of the coauthors, John C. Whitcomb, first became involved (as an evolutionist) in the study of origins at Princeton University in the summer and fall of 1942 in courses in historical geology and paleontology. The following year he became a Christian, and during the subsequent thirty-five years has developed an increasing interest in the question of ultimate origins in the light of the Book of Genesis. In 1957 a doctoral dissertation on the Biblical doctrine of the Flood was submitted to the faculty of Grace Theological Seminary, Winona Lake, Indiana (published in 1961 by the Presbyterian and Reformed Publishing Company under the title *The Genesis Flood: The Biblical Record and Its Scientific Implications*, co-authored with Henry M. Morris). In 1962 a presidential address was given at the Seventh General Meeting of the Midwestern Section of the Evangelical Theological Society in Chicago (published in 1963 as *The Origin of the Solar System* by the Presbyterian and Reformed Publishing Company). In addition to various articles in Bible dictionaries and encyclopedias, published works have included *The Early Earth* (Grand Rapids: Baker Book House, 1972) and *The World That Perished* (Grand Rapids: Baker Book House, 1973). A cassette course on *The Bible and Science* has been prepared, and a chronology chart covering the period from the Creation to Abraham (Winona Lake, Ind.: BMH Books, 1977). Since 1951, Professor Whitcomb has taught courses on such subjects as The Book of Genesis and Christian Apologetics at Grace Theological Seminary and as guest lecturer in dozens of other Christian schools of higher learning.

Donald B. DeYoung, the other coauthor, is a physicist, majoring in the areas of solid-state and nuclear science. He entered the physics field with the goal of making original contributions to the knowledge of the physical world. An early faith in Christ has been continually strengthened by his technical investigations. His graduate studies at Michigan Technological University (M.S., 1968) and Iowa State University (Ph.D., 1972) involved the area of Mössbauer Spectroscopy, a method of material research using resonant nuclear radiation. Dr. DeYoung has published scientific

articles in *The Journal of Chemistry and Physics of Solids,
The Journal of Chemical Physics,* and the *Creation Research
Society Quarterly.* His interest in astronomy has developed
into popular astronomy courses at Grace College, Winona
Lake, Indiana (sharing the same campus with Grace Theological Seminary), where he has taught for five years. His current
astronomy interests include creationism, astrophotography,
and binary star studies. On several occasions, lunar thin-
section mineral samples have been made available to him for
research and educational purpose. Dr. DeYoung has found
great challenge and satisfaction in sharing his background and
Christian testimony with students.

Contributing author George Mulfinger, Jr., (cf. Chapter V,
"Transient Lunar Phenomena") has enjoyed a broad background in the fields of science teaching, research, and scientific journalism. Professor of Physics and Physical Sciences at
Bob Jones University since 1965, he has coauthored a series
of science textbooks and has been a frequent contributor to
the publications of the Creation Research Society. He grad-
uated *summa cum laude* in chemistry from Syracuse University in 1953 and received the master of science degree in
physics from the same institution in 1962. He has done additional graduate work at Harvard and the University of
Georgia, and has received research grants from the National
Science Foundation and the Creation Research Society.

It is the sincere prayer of the authors that God may be
pleased to use this volume to strengthen His people everywhere as they rely on the full historicity and scientific accuracy of the Bible in its teachings concerning the origin of
the universe. We are convinced that it is only through a
proper understanding of God's Word that men can understand how all of reality fits together and finds its ultimate
interpretation. It is profoundly true of lunar science as of all
other sciences, that in Christ the Creator "are hid all the
treasures of wisdom and knowledge" (Col. 2:3).

John C. Whitcomb
Donald B. DeYoung
May 1, 1978

Acknowledgments

Not even a large team of qualified investigators can hope to discover complete answers to the major questions concerning the origin and nature of the moon. Converging and overlapping disciplines such as physics, chemistry, geology, mineralogy and mathematics are all vitally involved in lunar studies from an empirical standpoint. Above and beyond these are the theological disciplines of Biblical exegesis and hermeneutics for the purpose of unveiling God's special revelation in Scripture concerning the origin and purpose of the earth's great satellite.

It is in humble recognition of these facts that the present authors have submitted the manuscript of this volume in its various stages of development to a large number of specialists who have indicated particular interest in this project. We wish to acknowledge with genuine gratitude the suggestions and assistance of these men and to thank them for their sacrificial involvement and encouragement. The authors have endeavored to follow these suggestions either in correcting the manuscript or in answering more effectively the questions that were raised. Of course, we must emphasize the fact that full responsibility for the contents of this volume is being assumed by the authors. In fact, probably not a single reviewer would concur with *everything* we have written here. We solicit the help of yet others in eliminating remaining errors of fact or omission in anticipation of a second edition.

An asterisk (*) before the name of a reviewer indicates that he read more than one draft of the manuscript or provided special assistance.

Paul D. Ackerman, Ph.D., Professor of Psychology, Wichita State University

*Harold L. Armstrong, M.Sc., Professor of Physics, Queen's University, Kingston, Ontario

Greg L. Bahnsen, Th.M., Ph.D. (cand.), Assistant Professor of Apologetics, Reformed Theological Seminary

*Thomas G. Barnes, Ph.D., Professor of Physics, University of Texas at El Paso

Richard A. Bliss, M.A., Christian Heritage College and Institute for Creation Research

James Bone, B.S., Pilot, United Airlines, Chicago, Illinois

*Gerardus D. Bouw, Ph.D., Astronomer, East Cleveland, Ohio

James L. Boyer, Th.D., Professor of New Testament and Greek, Grace Theological Seminary

David R. Boylan, Ph.D., Dean, College of Engineering, Iowa State University

Robert H. Brown, Ph.D., Director of Geoscience Research Institute, Berrien Springs, Michigan

Larry Butler, Ph.D., Professor of Biochemistry, Purdue University

Charles A. Clough, M.A., Th.M., Pastor and Theologian, Lubbock, Texas

Gary Cohen, Th.D., President, Graham Bible College

John J. Davis, Th.D., Executive Vice President and Professor of Old Testament and Hebrew, Grace Theological Seminary

Vilas Deane, Ph.D., Associate Professor of Mathematics, Grace College

Paul Demmie, Ph.D., Idaho National Laboratory, Idaho Falls

Paul R. Elbert, Ph.D., Assistant Professor of Physics and Astronomy, California State University at Northridge

Donald L. Fowler, Th.M., Assistant Professor of Old Testament and Hebrew, Grace Theological Seminary

John M. Frame, M.Phil., Associate Professor of Apologetics and Systematic Theology, Westminster Theological Seminary

Bert Frye, M.S., Associate Professor of Astronomy and Geology, Cedarville College

Charles W. Harrison, Jr., Ph.D., Consulting Engineer, Director, General Electro-Magnetics, Albuquerque, New Mexico

Thomas H. Henderson, B.S., B.A. (physics and mathematics), Aerospace Engineer, Flight Simulation Division, NASA-JSC, Houston, Texas

Hilton Hinderliter, Ph.D., Assistant Professor of Physics, Penn State University, New Kensington Campus

Bruce Hrivnak, M.A., Astronomy Department, University of Pennsylvania

Jesse Humberd, Ph.D., Professor and Head, Division of Natural Science, Grace College

Robert D. Ibach, Jr., Th.M., M.L.S., Librarian and Assistant Professor of Old Testament and Archeology, Grace Theological Seminary and Grace College

R. Glen Ingalsbe, Electronics Technician, Kent, Washington

James B. Irwin, Astronaut, Apollo 15, Colorado Springs, Colorado

Richard Jeffreys, Ph.D., Professor of Biology, Grace College

Frank Jones, Ph.D., Professor of Mathematics, Rice University

Sherman P. Kanagy II, Ph.D. (astronomy), Visiting Lecturer in Physical Science, University of Illinois, Urbana, Illinois

Robert E. Kofahl, Ph.D., Physicist, Creation-Science Research Center, San Diego, California

Marvin L. Lubenow, M.S., Fort Collins, Colorado

Dowell E. Martz, Ph.D., Professor of Physics, Pacific Union College

John R. Meyer, Ph.D., Department of Physiology and Biophysics, University of Louisville

John A. Mitchell, M.A., B.D., Instructor in Astronomy, Mt. Clemens, Michigan

*John N. Moore, Ed.D., Professor of Natural Science, Michigan State University

*Henry M. Morris, Ph.D., Director, Institute for Creation Research; Academic Vice President, Christian Heritage College, El Cajon, California

Ellen Myers, Secretary, Bible Science Association, Wichita, Kansas

*Stuart E. Nevins, M.S., Visiting Professor of Geoscience, Christian Heritage College

*Robert C. Newman, Ph.D. (astrophysics), Associate Professor of New Testament, Biblical School of Theology, Hatfield, Pennsylvania

Philip G. Passon, Ph.D., Vice President, Development Departments, Servicemaster Industries, Inc., Downers Grove, Illinois

John G. Read, M.S.E.E., Aerospace Engineer, Culver City, California

*Larry G. Redekopp, Ph.D., Department of Aerospace Engineering, University of Southern California

William B. Rich, Ph.D., Associate Director of Educational Programs, Office of Public Affairs, NASA, Washington, D.C.

David J. Rodabaugh, Ph.D., Associate Professor of Mathematics, University of Missouri, Columbia, Missouri

Rousas J. Rushdoony, M.A., B.D., President, Chalcedon Foundation, Vallecito, California

Charles C. Ryrie, Ph.D., Th.D., Professor of Systematic Theology, Dallas Theological Seminary

Robert Sears, Ph.D., Professor of Physics and Astronomy, Austin Peay

State University, Clarksville, Tennessee

Wayne L. Slattery, Ph.D., Center for Astrophysics, Harvard College Observatory

Harold S. Slusher, M.S., Assistant Professor of Geophysics, University of Texas at El Paso

Charles R. Smith, Th.D., Professor of Christian Theology and Greek, Grace Theological Seminary

*Michael S. Spence, A.B., Student, Grace Theological Seminary

Paul M. Steidl, M.S. (astronomy), Computer Programmer, Alderwood Manor, Washington

Peter M. Thompson, B.S. (geology), Officer, U.S. Marine Corps

Lawrence E. Turner, Jr., Ph.D., Associate Professor of Physics and Computer Science, Pacific Union College

Stan Udd, Th.M., Instructor in Bible and Theology, Calvary Bible College

A. James Wagner, B.A., M.S., Meteorologist, National Weather Service, Camp Spring, Maryland

David C. Whitcomb, B.S., pre-med., N.I.H., Bethesda, Maryland

Stan Wineland, M.S., Assistant Professor of Physics and Director of the Newhard Planetarium, Findlay College

William R. Winkler, B.S., M.S., Scientific writer; editor, Environmental Data Service, National Oceanic and Atmospheric Administration, Washington, D.C.

*Donald G. York, Ph.D., Research Astronomer, Princeton University

Larry Zavodney, M.S. (mechanical engineering), Diamond, Ohio

George J. Zemek, Jr., Th.M., Instructor in Greek, Grace Theological Seminary

It has been our pleasure to work with Rev. Charles W. Turner, Executive Editor and General Manager of the Brethren Missionary Herald Company, Winona Lake, Indiana, and Mr. Kenneth E. Herman, Assistant to the General Manager. Their encouragement and helpful counsel in the preparation of this volume have been deeply appreciated by the authors.

Finally, special thanks are due to Mrs. Betty Vulgamore and Mrs. Cathy Miller, secretaries of the faculty, Grace Theological Seminary, who typed the manuscript; and to the writers' families, without whose patience and prayerful encouragement this project could never have been completed.

John C. Whitcomb Donald B. DeYoung
Grace Theological Seminary Grace College

May 1, 1978   Winona Lake, Indiana

CHAPTER I

Lunar Exploration

"In the beginning God created the heaven and the earth" (Gen. 1:1)

(This opening verse of the Bible and the nine verses that follow it were read from the vicinity of the moon on Christmas Eve, 1968, by astronauts William Anders, James Lovell, and Frank Borman. These were the first human beings to escape the earth's gravitational field and to orbit the moon.)

Hundreds of military, communications, and remote sensing satellites have orbited our world, beginning with the 1957 Russian Sputnik. Several of these satellites can be seen each clear evening as wandering points of reflected sunlight. Each, however, is artificial and of temporary use only. Taken altogether, they cannot compare in grandeur or in purpose with our single natural satellite, the moon. From the beginning man has enjoyed the divinely appointed nocturnal illumination of the moon, as well as its regular cycles, so useful for timekeeping.

This publication has been prepared to explore the creation, purpose, nature and destiny of our moon from a Biblical and scientific perspective. The current space program has been a disappointment to many, with its emphasis on a futile search for evolutionary evidence contrary to a Biblical creation. This is true of the analysis of space radiation for intelligent signals, of the Viking quest for life on Mars, and it is also true of the Apollo search for a natural model of lunar origin. All too often the conclusions of the secular scientific establishment are disseminated as final authority. However, with the continuous flow of new and more refined data, the natural laws and models discovered by scientists never really arrive at a final, perfect state. This work is presented with a firm commitment to the authority of Scripture over the theories of men. God's Word provides a final authority which applies to the created universe as well as to every human need.

The moon is actually less than one percent as far away as the earth's next nearest major neighbor, the brilliant planet Venus. The earth-moon distance averages nearly 384,000 kilometers (239,000 miles). Equivalent to ten trips around the world, this is less distance than some automobiles have traveled! One may conceive of this proximity as follows:

> Twenty-nine earths placed side by side in a straight line would form a suspension bridge long enough to connect the two worlds. This insignificant distance is hardly worthy to be called astronomical. A light signal would take just over a second to cover this distance, which is a mere 400th of the distance to the sun. To reach any of the stars, the distance to the moon would have to be repeated nearly 100 million times.[1]

Though the moon is tantalizingly close, it remains mysterious, and numerous basic and difficult questions naturally arise. "What is the moon made of?" "Where did it come from?" Technology has increasingly provided insight for these and other uncertainties in the sky. Galileo, who viewed the moon through his telescope in 1609, was among the first scientists to begin to find answers to such questions. He was surprised by the rugged landscape, and suggested that the dark areas were literal oceans. In *The Starry Messenger*, published in 1610, Galileo described the moon's surface as uneven, rough, and packed with cavities and protrusions.[2] Since 1948 the Mount Palomar telescope near San Diego, one hundred times larger in diameter than that of Galileo, has

---

[1] Camille Flammarion, *The Flammarion Book of Astronomy* (New York: Simon and Schuster, 1964), p. 85.

[2] G. Galileo, *The Starry Messenger*, translated by Stillman Drake (Garden City, New York: Doubleday Anchor Books, 1957), p. 31.

remained unsurpassed in earth-based studies of the visible part of the moon and the heavens beyond.[3] It was not until 1959, however, that the Soviet spacecraft Luna 3 provided the first photographs of the moon's hidden or far side. Fanciful speculation about wonders on the far side were ended by these photographs of a heavily cratered, mountainous surface. (See Table I-1 for a compilation of lunar exploration events.) These many early efforts provided not only a foundation but a strong impetus for the culminating Apollo program of lunar exploration, which climaxed with Neil Armstrong's dramatic footsteps on the Sea of Tranquillity in 1969, and concluded with Apollo 17 in December, 1972.

After remaining beyond man's grasp through history, the moon in 12 short years experienced landings by 21 Soviet and 13 American unmanned space craft, and 6 manned Apollo crews. A dozen astronauts traveled on the lunar surface, covering 60 miles and conducting 50 major experiments. At its peak the American space program supported 400,000 technical jobs. (See chart next page.)

Moon research today has not been abandoned, even though no further manned American craft are scheduled to make the trip this century. On the contrary the study has just begun. Both the number and complexity of questions in any scientific field multiplies with time, and lunar research has been no exception. Instruments left behind continued to monitor moonquakes until all Apollo experiments were terminated in October 1977, for other priorities. Automated transmitters on the moon still continue their beacon, refining the study of celestial mechanics as the moon responds to the forces of earth and solar gravity. The 382 kilograms (843 pounds) of lunar material returned by Apollo teams will be investigated by hundreds of scientists for years to come. Strategic computer terminals across the United States have been established to provide quick access to the accumulating

---

[3] In 1976 a 6 meter (236 inch) reflecting telescope began operating in the Soviet Union's northern Caucasus Mountains. Its mirror has a 39 percent larger area than the 5 meter (200 inch) Palomar reflector. Atmospheric disturbance of starlight presently rules out one-piece ground telescopes of still larger size. Future instruments such as the 2.4 meter (95 inch) American Space Telescope are planned for artificial satellites or locations on the moon to avoid atmospheric distortion of images.

Date	Event	Credit
1609	Telescopic Sketch of the Moon	Galileo
1647	Accurate Lunar Map	Helvetius
1840	First Detailed Lunar Photographs	(Various Sources)
October, 1959	Photographs of Moon's Backside	Soviet Union
July 31, 1964	Ranger VII Close-up Lunar Images	United States
February, 1966	Unmanned Soft Landing	Soviet Union
Dec. 21-27, 1968	Apollo 8 (Lunar Orbit)	F. Borman, W. Anders, J. Lovell
July 16-24, 1969	Apollo 11 (Mare Tranquillitatis)	N. Armstrong, E. Aldrin, M. Collins
Nov. 14-24, 1969	Apollo 12 (Oceanus Procellarum)	C. Conrad, A. Bean, R. Gordon
September, 1970	Luna 16 (Mare Fecunditatis)	Soviet Union
Jan. 31-Feb. 9, 1971	Apollo 14 (Fra Mauro Hills)	A. Shephard, E. Mitchell, S. Roosa
July 26-Aug. 7, 1971	Apollo 15 (Apennine Mountains)	D. Scott, J. Irwin, A. Worden
February, 1972	Luna 20 (Apollonius Highlands)	Soviet Union
April 16-27, 1972	Apollo 16 (Descartes Highlands)	J. Young, C. Duke, T. Mattingly
Dec. 7-19, 1972	Apollo 17 (Taurus-Littrow Valley)	E. Cernan, H. Schmitt, R. Evans

Table I-1   Summary of Major Lunar Exploration Events

The Apollo/Saturn V space vehicle lifts Apollo 15 astronauts toward the moon on July 26, 1971. The 363-foot rocket generates 7.5 million pounds of thrust. (NASA)

The 1,000-foot (300 meter) diameter radio telescope of the Arecibo Observatory in Puerto Rico, operated by Cornell University. This instrument allows scientists to study solar system objects with radar, as far out as the planet Saturn. (NASA)

The 200-inch (5 meter) Hale Telescope dome on Palomar Mountain. (Photo by Dr. Lester E. Pifer)

A full moon photographed from the Apollo 11 spacecraft at a distance of 10,000 miles from earth. This near-side picture is similar to the crude sketches that Galileo drew. (NASA)

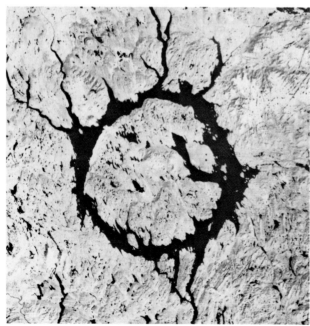

Viewed from an altitude of 500 miles, the circular Manicauagan structure in Quebec, Canada, appears to be a deeply eroded crater. The central region (surrounded by the black-appearing waters of a reservoir) still preserves rocks that were shattered and melted by the force of the impact. (NASA)

The Apollo 11 lunar module ascent stage during rendezvous with the command service module. The earth rises above the lunar horizon. (NASA)

lunar data stored in government facilities in Maryland.[4] It was estimated in February 1976, that 30,000 pages of scientific literature had already been generated by the intense lunar study. An international journal concerned exclusively with lunar studies is flourishing.[5]

What is the value of this Apollo project to mankind? The National Aeronautics and Space Administration (NASA) budget during the Apollo years and Mercury-Gemini preparations totaled $25-$30 billion, more than $100 spent for every American citizen! In addition to the enormous financial investment, it must be remembered that three astronauts were tragically lost in the Apollo I test spacecraft fire on January 27, 1967: Virgil Grissom, Edwin White, and Roger Chaffee. Many people are critical of such sacrifices made to satisfy dreams of exploration when human needs abound on planet earth.

While some had predicted that lunar exploration would signal the beginning of lunar militarization or colonization, such predictions were unfounded and have not materialized. The moon is much too distant to be a strategic military base, and its inhospitable environment does not encourage human settlement.

On the positive side of the Apollo effort is its success in providing the United States with a capability of living and working in a space environment. Apollo has thereby provided a vast storehouse of new knowledge which is being utilized in technological improvements, sometimes called "spin-offs." These include safer and more efficient aircraft, new vistas in medical science, agricultural breakthroughs, and a host of other applications to everyday living.[6] Other areas of fundamental space research remain less practical for the present.

---

[4]Descriptive brochures and a list of terminal locations are available from the National Aeronautics and Space Administration, Washington, D.C. 20546.

[5]*The Moon:* An International Journal of Lunar Studies (Dordrecht, Holland: D. Reidel Publishing Co., 1969-).

[6]James J. Haggerty, *Spinoff 1977,* Available from the Superintendent of Documents, U.S. Government Printing Office, Washington, D.C. 20402. See also Neil P. Ruzic, *Spinoff 1976: A Bicentennial Report* (Washington, D.C.: National Aeronautics and Space Administration Technology Utilization Office, 1976); and F. I. Ordway, *Dividends From Space* (New York: Thomas Y. Crowell Co., 1971).

For example, studies in the identification of lunar minerals and the measure of celestial movements are mainly of academic interest. Other topics such as the moon's origin, its influence on earth, and its hospitality to life continue to arouse curiosity and receive emphasis in the popular literature. It is in these areas, particularly the subject of origins, that the layman can appreciate the advances of exploration and research. Unfortunately, with Apollo as with many other scientific endeavors, much of the public information has a strongly humanistic, evolutionary basis. The whole truth is seldom told concerning physical evidence and conclusions, such as discordant ages, as this study will show in the case of Apollo. Natural science is by definition experimental, and can therefore give neither absolute nor timeless answers to the questions it examines. An endless cycle of new data and new theories replace the old. Some changes are narrow refinements; others are revolutionary.

Amidst the glory of scientific achievements such as Apollo, it is well to remember the prophet Jeremiah's counsel that creation is in the end inscrutable: the heavens can never be fully measured, nor the foundations of the earth searched out (Jer. 31:37). Only the special revelation of God through His Son and His Word provides final answers in matters of origins. God was, after all, the only One there! Thus the Christian can be confident that *valid* scientific conclusions will be in accord with God's creation message, and the Apollo mission to the moon is no exception. All findings have been found to be in harmony with the teachings of Scripture concerning the moon, for the earth's satellite is apparently designed for the specific purpose of meeting human needs and is recent in origin (Gen. 1:16).

CHAPTER II

Naturalistic Theories
of the Moon's Origin

*"A blind acceptance of [evolutionary theory] may in
fact be the closing of our eyes to as yet undiscovered
factors which may remain undiscovered for many years
if we believe that the answer has already been found. It
seems at times as if many of our modern writers on
evolution have had their views by some sort of revelation . . . It is premature, not to say arrogant, on our part
if we make any dogmatic assertion as to the mode of
evolution."*

G. A. Kerkut
*(*Implications of Evolution, *p. 155)*

Even a cursory review of current scientific theories about
the moon's beginning will convince the reader that there is no
consensus on origin. In fact, no naturalistic lunar explanation
conforms to the established laws of celestial mechanics. The
standard complex and intertwined theories nevertheless continue to be offered as probable because there are no alternative explanations apart from supernatural creation, which is
true for the earth as well as for the moon. One can hear the
searchers ask in exasperation: "Why is it that the body [the
moon] with the most mysterious origin in the Solar System
dominates the night sky?"[1] Pre-Apollo hopes were strong that

---

[1] A. L. Hammond, "Exploring the Solar System (III): Whence the Moon?"
Science, 186:4167 (December 6, 1974), 911.

one origin approach would emerge from lunar exploration as credible, but it did not happen. On the contrary, the traditional origin theories have only been further discounted. *Time* magazine concludes:

> Post-Apollo studies have shown that there is no overall explanation for the origin of the planets or their moons. Some scientists feel that they may never completely learn the origin and history of the earth's immediate neighbor.[2]

A state of confusion is shown by divergent statements from the scientists privileged to publish Apollo data. S. R. Taylor of the Lunar Science Institute said: "The Apollo missions have revolutionized our knowledge of the moon. This has rendered all previous theories obsolete, a situation that has had many parallels in science and history."[3] University researchers R. A. Pacer and W. D. Ehmann wrote: "None of these theories [on lunar origin] can be definitely ruled out on the basis of information gained from Apollo missions."[4] A summary of the Third Lunar Science Conference in 1972 stated:

> The most profound question of all is the origin of the moon, or more properly the earth-moon system. Apollo science has eliminated the once-popular hypothesis that the moon was captured recently (1 to 2 billion years ago), but beyond this the question remains unanswered.[5]

Three major naturalistic (non-supernatural) theories have been developed to explain the moon's presence. These theories postulate, respectively, (1) a lunar origin by fission (splitting off) from the earth; (2) capture by the earth's gravitational field; or (3) condensation from nebulous gas and dust at the same time the earth was formed. These three views are here considered, in order to show their unsatisfactory answer to the origin question.

---

[2]"The New Moon," *Time*, 103:114 (April 8, 1974), 52.

[3]S. R. Taylor, *Lunar Science: A Post-Apollo View* (New York: Pergamon Press, Inc., 1975), p. 306.

[4]R. A. Pacer and W. D. Ehmann, "The Apollo Missions and the Chemistry of the Moon," *Journal of Chemical Education*, 52:6 (June, 1975), 350.

[5]"Third Lunar Science Conference," *Science*, 176:4038 (June 2, 1972), 981.

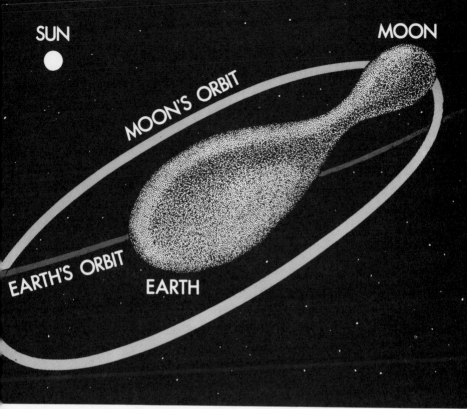

SUN

MOON

MOON'S ORBIT

EARTH'S ORBIT

EARTH

Figure II-1. The fission ("daughter") theory contends that the earth initially rotated very rapidly, and tore apart. The smaller mass entered into orbit as the moon.

Sketch by Arthur Davis

(1) THE FISSION OR BREAK-AWAY THEORY

The British astronomer and mathematician George H. Darwin (1845-1912), son of Charles Darwin, spent his life studying the earth's tides and their effect on earth history. In 1879 George Darwin announced his conclusion that the moon had originally split off from the earth.[6] The explanation is variously called a "daughter," "fission," or "break-away" theory. Darwin proposed that the sun caused a substantial tidal bulge on a rapidly spinning molten earth early in its history. The earth subsequently became lopsided and eventually hot

[6]G. H. Darwin, "On the bodily tides of viscous and semi-elastic spheroids, and on the ocean tides on a yielding nucleus," *Philosophical Transactions of the Royal Society of London*, 170 (1879), 1-35. See also G. H. Darwin, "On the secular changes in the orbit of a satellite revolving around a tidally disturbed planet," *Philosophical Transactions of the Royal Society of London*, 171 (1880), 713-891.

chunks separated, similar to what happens when mud is thrown from a spinning bicycle wheel. The pieces then accumulated in orbit and formed our moon. The Pacific Ocean was later proposed as the "scar" left behind by the lunar mass.

Even though strongly criticized by the British astronomer Harold Jeffries in 1929,[7] the fission theory continues to remain under serious consideration. However, the problems are many. The most serious problem is that an earth rotation period of about 2.6 hours is necessary for the proposed resonant rotational instability.[8] This is considered to be an impossible rotation rate even by evolutionary extrapolation. A rotation period of as little as 5 hours, with the moon initially a few earth radii away, would cause tides several kilometers in height, an extensive melting of the earth's mantle, and the evaporation of much ocean water.[9] R. B. Baldwin describes the problems involved in tidally braking the earth's spin as follows:

> If the rotation of the earth had been slowed by tidal friction from 4 to 24 hours, there would have been an energy dissipation of 1.2×10^{10} ergs/gram within the earth. Such an energy release is sufficient to raise the temperature of the entire earth by 1000°C.[10]

In addition, any theory requiring an originally rapidly spinning earth must account for the disappearance of one-half of its initial spin angular momentum. The spin angular momentum of the earth is a measure of its rotational motion, and is proportional to the product of its mass, radius, and equatorial speed. The present angular momentum of the earth-moon

---

[7] H. Jeffries, *The Earth* (Cambridge: Cambridge University Press, 1929).

[8] P. Goldreich, "Tides and the Earth-Moon System," *Scientific American*, 226:4 (April, 1972), 51.

[9] *Ibid*, p. 50. The vaporization of the entire world oceans is predicted by W. H. Munk, "Once again - tidal friction," *Quarterly Journal of the Royal Astronomical Society*, 9 (1968), 352-375.

[10] R. B. Baldwin, *A Fundamental Survey of the Moon* (New York: McGraw-Hill Book Company, Inc., 1965), pp. 42-43. On the uniformitarian basis of periodicities of Precambrian stromatolites, a close approach of the earth and moon about 2.85 billion years ago is concluded by D. L. Turcotte, J. L. Cisne, and J. C. Nordmann, "On the Evolution of the Lunar Orbit," *Icarus*, 30 (1977), 254-266. However, evidence is lacking for simultaneous earth heating.

system is only half of that necessary for an original rotational instability. Attempts have been made to explain the non-conservation of angular momentum by meteorological effects and processes within the earth, but no adequate mechanism has been found.[11]

Lunar laser-ranging experiments regularly measure the earth-moon distance to within a few centimeters,[12] and it is noticed that the separation is presently increasing by several centimeters per year. The *rate* of separation, however, based on uniformitarian extrapolation, is decreasing; the rate of separation is assumed to have been greater in the past.[13] At the same time the length of the earth's day is increasing by about 0.002 seconds per century, both effects being due to the tidal forces between the earth and the moon. Even this slow rate of tidal dissipation of energy is extremely large when compared with the fission theory and its assumed 4-billion-year earth history. If the present rate of earth-moon separation is extrapolated backward in time, the moon would have been "very near" the earth less than two billion years ago! Baldwin explains the situation this way:

> Jeffrey's (sic) early studies of the effects of tidal friction yielded a rough age of the Moon of 4 billion years. Recently, however, Munk and MacDonald have interpreted the observations to indicate that tidal friction is a more important force than had been realized and that it would have taken not more than 1.78 billion years for tidal friction to drive the Moon outward to its present distance from any possible minimum distance. This period of time is so short, compared with the age of the earth, that serious doubts have been cast upon most proposed origins and histories of the moon.[14]

Allen Hammond, using slightly different initial conditions, concludes that the current rate of separation of the earth-

---

[11]W. H. Munk and G. J. F. MacDonald, *The Rotation of the Earth* (London: Cambridge University Press, 1960).

[12]Taylor, *Lunar Science: A Post-Apollo View*, p. 3.

[13]G. J. F. MacDonald, "Origin of the Moon: Dynamical Considerations," *The Earth-Moon System*, edited by B. G. Marsden and A. G. W. Cameron [Proceedings of an International Conference, January 20-21, 1964] (New York: Plenum Press, 1966), p. 185.

[14]Baldwin, *A Fundamental Survey of the Moon*, p. 40.

moon system implies an initial separation less than one billion years ago.[15] This is long after the date given to the youngest rocks found on the moon (see p. 92). There is an obvious error with either the fission theory or the assumed long ages. As will be seen, it is more likely that both ideas are false.

There is an unanswered question as to whether the majority of tidal friction takes place in the earth's interior or in its surface waters. If the surface is important, extrapolated tidal heating in the earth's history conflicts directly with the geologic time scale:

> If tidal friction takes place largely in the solid Earth, then one could feel reasonably safe in extrapolating the present value [of the rate of change of the earth's rotation] into the geologic past (allowing, of course, for the increased torque with the diminished distance from the Moon) ... If, on the other hand, tidal friction is largely associated with shallow seas, the extrapolation even for 10,000 years is hazardous [to the fission model]. The tidal heating would have become an essential element in the thermal history of the Earth, and things might have been quite different from what they are now.[16]

All investigations thus far suggest that most, if not all, of the earth's tidal dissipation does indeed take place in the shallow seas.[17]

Another difficulty with the fission model is that such an occurrence of a tidal separation could not even lead to a permanent moon. Gordon MacDonald of the Institute of Geophysics and Planetary Physics, University of California, concludes from his calculations:

> A mass thrown off by the Earth would initially have an angular velocity less than the rotational velocity of Earth. Tidal action of the thrown-off Moon, traveling in such a retrograde orbit [an apparent westward motion], would tend to bring the Moon back to Earth. A thrown-off mass would never escape.[18]

---

[15]Hammond, "Exploring the Solar System (III): Whence the Moon?" 911.

[16]W. H. Munk, "Variation of the Earth's Rotation in Historical Time," *The Earth-Moon System*, 67-69.

[17]Goldreich, "Tides and the Earth-Moon System," 48.

[18]G. J. F. MacDonald, "Origin of the Moon: Dynamical Considerations," *The*

Additional problems of a terrestrial origin for the moon include the extreme differences in water content, density, and element abundances of the two bodies. One reason for originally proposing the fission theory was to explain the low density of the moon, only 60 percent that of earth. This density difference is apparently due to an iron deficiency on the moon, compared with earth.[19] It was attractive to think of the moon's material as having come from the outer parts of the earth which are less dense than the core. However, the other physical differences now known between the earth and moon are a major argument against the fission theory. Hammond, in reviewing earth-moon similarities, concludes that "some substantial chemical differences remain."[20] The earth and moon appear to be very different in many respects, rather than common fission products! Also, the moon's orbit is found to be inclined to the earth's equator by a cyclically varying angle between 28½° and 18½°. Ejection from earth would have certainly inserted the moon into an equatorial plane, since rotational speed is greatest at the earth's equator.

Finally, a moon thrown off by the earth must pass through "Roche's Limit." This is the minimum distance at which a satellite can withstand the gravitational forces exerted on it by its planet without disintegrating. The rings of Saturn are particles which are closer to the planet than the distance at which a large solid body can exist—that is, they are within Roche's limit. For the earth-moon combination, the breakup distance (i.e. Roche's limit) is about 2.89 times the earth's radius, or 18,500 kilometers (11,500 miles) from the earth's center.[21] The moon, however, is stable, being 21 times as far

Earth-Moon System, p. 200; G. J. F. MacDonald, "Tidal Friction," *Reviews of Geophysics*, 2 (1964), 467-541. Another argument for an initial retrograde orbit of the moon is presented by H. Gerstenkorn, "Uber Gezeitenreibung beim Zweikorperproblem," *Zeitschrift für Astrophysik*, 36 (1955), 245-274.

[19]W. M. Kaula and A. W. Harris, "Dynamics of Lunar Origin and Orbital Evolution," *Reviews of Geophysics and Space Science*, 13:2 (May, 1975), 363-371. This article contains an exhaustive list of 80 references on lunar origin papers.

[20]A. L. Hammond, "The Moon: Not so different from Earth after all," *Science*, 192:4242 (May 28, 1976), 875.

[21]In 1850 Edouard Roche (1820-1883 A.D.) demonstrated that a satellite

away as this breakup distance.

Many modifications of the fission model in recent years have been attempted to salvage the basic idea of an earth origin for the moon. An assortment of centrifugal, tidal, volcanic, and vibrational influences have been marshalled to provide a fission mechanism. However, all attempts remain without a solid physical basis. The conclusion was well stated by an official NASA summary of the Apollo project:

> Where did the Moon come from? This is the puzzler of them all. One theory is that the Moon came from Earth, possibly wrenched from what is now the Pacific Ocean. Most students of the subject reject this possibility because it proves very difficult to explain all the steps that must have occurred to bring this about.[22]

(2) THE CAPTURE THEORY

The earth's gravitational force maintains the moon in its stable orbit. In fact, gravity is the attractive force that holds all massive objects together in the physical universe. It is this force that is appealed to in the theory that the moon was trapped by the earth as it wandered through the solar system in its own solar orbit. Of course, this suggestion that the earth's gravitational field captured the moon simply moves the lunar origin problem to somewhere else. Even apart from that unsolved problem, however, the capture process remains extremely implausible, as Goldreich states:

> Problems with capture theories are in large part due to the implausibility of the capture process, which requires the velocity

---

would be torn apart by tidal forces if it approached its planet closer than a distance d, where

$$d = 2.4554 \left(\frac{\rho_p}{\rho_m}\right)^{1/3} R.$$

The quantities ρ_p and ρ_m are respectively the planet and satellite densities, and R is the radius of the planet. If the densities ρ_p and ρ_m are the same, then

$$d = 2.4554 \, R.$$

For the earth and its moon $\rho_p = 5.52$ gms/cm^3 and $\rho_m = 3.34$ gms/cm^3. Hence

$$d = 2.4554 \left(\frac{5.52}{3.34}\right)^{1/3} R.$$
$$= 2.89 \, R$$

[22] E. M. Cortright, *Apollo Expeditions to the Moon* (Washington, D.C.: NASA, 1975), p. 301.

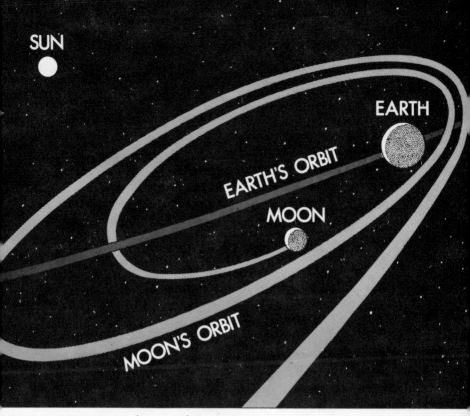

Figure II-2. The capture ("spouse") theory holds that the moon came from a separate solar orbit. Sweeping too close to the earth, it was captured by the earth's gravitational attraction and locked into orbit.

Sketch by Arthur Davis

of an initially unbound moon to be reduced by some kind of dissipation process.[23]

Hammond has given some details on the constraints for lunar capture.[24] He explains that in order for the moon to be captured intact, its speed could not exceed 40 meters/second. If its speed exceeded 2,500 meters/second, it would be diverted into a new heliocentric orbit, not captured at all. In between these two extremes it would have broken into particles and distributed itself into Saturn-type rings. Jupiter, which is more massive than all other planets of the solar system combined, has apparently altered the orbits of a whole family of comets, giving them short 5-10 year revolv-

---

[23] Goldreich, "Tides and the Earth-Moon System," 51.

[24] Hammond, "Exploring the Solar System (III): Whence the Moon?" 912.

ing periods about the sun.[25] Many other comets are ejected from the solar system when planetary perturbations change comet orbits from ellipses to hyperbolas. However, no comets have been observed changing into planetary satellites by a capture process. There is simply no known means by which the moon's velocity could be largely dissipated on a single pass. For example, energy-losses from tidal friction and particle collisions compared to total kinetic energy are negligible in a single encounter. Extreme attempts have been called upon to bolster the capture idea. Baldwin comments on one:

> It has even been suggested that the earth [originally] had a distant, small moon. Our present moon, in one of its passages near the earth, collided with the little moon, thus losing orbital energy and permitting capture by the earth. The objection is the usual one. Too much heat would be involved. The colliding bodies would have been largely vaporized. No one has yet devised an acceptable mechanism for the capture of the moon by the earth.[26]

Recent calculations have shown that the moon's orbit may be unstable in the long run.[27] The study concludes that since the *earth's* orbital eccentricity (i.e., the deviation from a circular orbit) exceeds a calculated critical value, it is possible for the moon eventually to escape and become a planet. Also, it is said to be dynamically possible that in the remote past, the moon originated as a planet and then was captured by the earth, but this assumes a long time scale, both in the past and in the future. The calculations referred to, however, do not include tidal effects or the nonsphericity of the earth's shape. Hence, this recent conclusion that lunar capture is possible is subject to challenge, as are all natural origin theories.

Any theory for the origin of the moon must explain the *moon's* orbital eccentricity. The elliptical orbit of the moon is not eccentric enough (0.055) to make a capture origin

[25] George Abell, *Exploration of the Universe* (New York: Holt, Rinehart and Winston, 1969), p. 339.

[26] Baldwin, *A Fundamental Survey of the Moon*, p. 42.

[27] V. Szebehely and R. McKinzie, "Stability of the Sun-Earth-Moon System," *Astronomical Journal*, 82, (1977), 303.

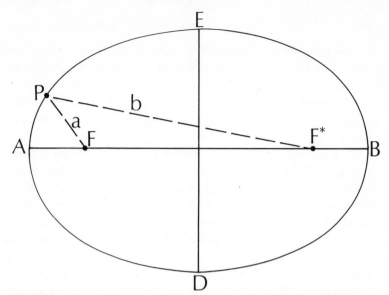

Figure II-3. The ellipse is the shape of all satellite orbits. The elements of an ellipse are shown: major axis (AB), minor axis (ED), and foci (F and F*). The sum of the distances of any point P on the ellipse to the two foci is always a constant (a + b is constant). For the earth-moon system, the earth would be located at one of the focal points. The moon would be at the point P.

likely (Fig. II-3). Such an origin, if possible at all, would probably produce a comet-like, greatly elongated orbit. Thomas Gold describes the situation:

> It is probable that if the Moon were captured by the Earth the eccentricity would initially have been much larger [than the present value]. The many-moon hypothesis . . . suffers from the same difficulty, for it seems unlikely that the largest protomoon could have an almost circular orbit after it had swept up all the smaller bodies.[28]

As with the fission model, lunar capture is recognized by many astronomers as a doubtful explanation. Even if a credible mechanism is eventually found to capture the moon, this will not demand that the moon *was* captured.

(3) THE NEBULAR OR CONDENSATION THEORY

The "nebular," "sister," or "condensation" theory calls

---

[28]T. Gold, "Long-Term Stability of the Earth-Moon System," *The Earth-Moon System*, 94.

for an independent accretion growth of earth and moon from dust and gases in the same region of space. Meteorites are considered to be similar aggregates of condensed material. Formulated by Gerard P. Kuiper and C. F. von Weizsäcker around 1951, this theory is favored by most astronomers today.

A crucial problem with this view is the precise balance needed during the proposed buildup of the proto-earth and proto-moon. Initially both bodies would have had small gravitational attractions. The orbital velocity of the early moon would necessarily have been small in order for the gravitational force to provide a sufficient inward centripetal binding force. As the objects grew in size, both the gravitational attraction and the orbital velocity must have increased very precisely to avoid lunar escape or a collision with the

Figure II-4. In the nebular ("sister") hypothesis, the earth and moon formed at the same time from a vast cloud of cosmic matter that condensed into the objects of the solar system.

Sketch by Arthur Davis

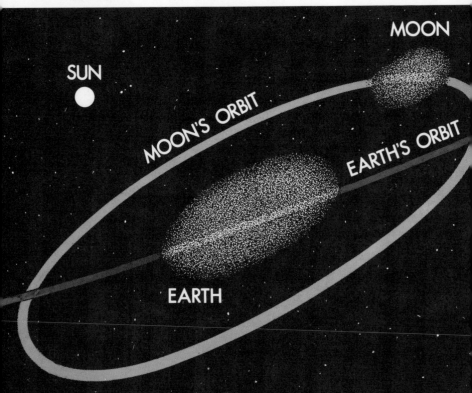

earth.[29]

One must be fully aware of the fundamental assumption of the nebular theory, that gases in space will condense to form initially a nebula and eventually a dense body. This assumption exists in general for star formation as well as for the lunar nebular hypothesis:

> Contemporary opinion on star formation holds that objects called protostars are formed as condensations from intersteller gas. This condensation process is very difficult theoretically, and no essential theoretical understanding can be claimed; in fact, some theoretical evidence argues strongly against the possibility of star formation. However, we know that stars exist, and we must do our best to account for them.[30]

Laboratory experiments to form contracting gas clouds face a *scaling* problem. Since gravity is an inverse square law force without canceling effects (as electromagnetism has), the total gravitational force at the 'surface' of a spherical gas cloud of fixed density and temperature will *increase* linearly with the radius. Therefore for *any* given temperature and density of gas, there *is* a maximum stable size, beyond which the cloud will contract irreversibly (as the density will increase faster than the temperature will rise).

Recent calculations have shown that interstellar gas clouds *may* gravitationally collapse if the unknown retarding forces of gas rotation, magnetic field, and gas turbulence are small.[31] The masses considered are all much greater than that of the sun. A critical factor is the initial density of the gas cloud, which must be high. Fanciful theories to create special starting conditions have not been convincing. A chance jostling of the gas cloud or a shock wave have been proposed to provide the initial mass density needed to overcome the out-

[29] G. J. F. MacDonald, "Origin of the Moon: Dynamical Considerations," in *The Earth-Moon System*, p. 202.

[30] J. C. Brandt, *The Sun and Stars* (New York: McGraw-Hill Book Co., Inc., 1966), p. 111. Contradictions of the nebular theory with the laws of thermodynamics are discussed by G. Mulfinger, "Critique of Stellar Evolution," *CRSQ*, 7:1 (June, 1970), 7-24.

[31] R. L. Dickman, "Bok Globules," *Scientific American*, 236:6 (June, 1977), 66-81. The physical details of gravitational collapse of interstellar gas are described by R. C. Newman and H. J. Eckelmann, Jr., *Genesis One and the Origin of the Earth* (Downers Grove, Illinois: InterVarsity Press, 1977), pp. 44-51.

ward gas pressure by gravitational attraction. An initial abundance of gravel-size particles have also been postulated to increase the density. However, this proposal begins to resemble the already-existing belt of asteroid particles beyond the orbit of Mars, which certainly shows no tendency to condense into a single solid. One begins to wonder how such naturalistic theories can endure in the face of such enormous problems.

If the origin of the moon actually involved natural condensation from a whirlpool of matter simultaneously with the earth's formation, then this explanation might well be extended throughout the solar system. There should be a harmony of motion, with objects rotating (motion of the object about its axis), and revolving in their orbits in the same direction. However, it is found that several of the solar system moons revolve in a direction opposite to the rotation direction of their parent planets. The first such abnormality was discovered in 1898, when the satellite Phoebe, of Saturn, was found to have this *retrograde* motion. Since that time, four outer moons of Jupiter and one of Neptune have been discovered to be revolving about their parent planet in the "wrong" direction. The five moons of Uranus also show retrograde motion, although this is basically a technicality since Uranus' axis of rotation is tilted almost into the plane of the solar system. What could have caused several of the solar system moons to establish retrograde orbits? Indeed, what could have caused the other objects to revolve in a "forward" direction? Natural origin theories cannot properly explain these complex irregularities. It is simply impossible to account for all the details of creation when the Creator is neglected.

Other problems for the earth-moon nebular formation theory include the marked differences in element abundance and density of the earth and moon, problems already noted in the fission theory. A natural origin from similar materials in the same region of space should result in similar properties.

CONCLUSION

The lack of any sufficient natural origin theory for the

Physical Evidence	Origin Theories		
	Fission	Capture	Condensation
Angular Momentum (L).	Initial *large* L is required for earth. L has not been conserved for the earth-moon system (50 percent loss).	Present L for the earth-moon system is large compared to the other planet-satellite systems.	
18½°-28½° inclination of moon's orbit to earth's equator.	No inclination is expected since moon would leave equatorial region.	A large inclination is expected from a random encounter.	There is no plausible explanation for nebula condensation.
Laws of Dynamics.	Origin is unstable. Moon would be pulled back to earth. Present rate of moon deceleration brings it back to earth in less than 2 billion years. Moon must escape the destructive Roche limit.	Capture is an extremely improbable event. Moon speed must be somehow reduced during the earth encounter. Capture theory does not explain the moon's initial origin.	There is an extremely improbable balance of moon position and speed during mass buildup, to maintain a stable orbit.
Density, chemistry, minerology differences between earth and moon.	Expect closer similarity.	No other location in the solar system completely explains the lunar composition.	Expect closer similarity.
Temperature.	There is a lack of evidence for an expected 1000°C heating of the earth by tidal energy dissipation.		

Table II-1. Summary of Major Problems of Naturalistic Lunar Origin Theories.
(See text for explanation)

moon is readily admitted by experts. One report of the seventh annual Lunar Science Conference began: "No Conference, such as the one at NASA's Johnson Space Center in Houston two weeks ago [March, 1976], would be complete without a new theory of the origin of the Moon."[32] Dr. Harold Urey, a Nobel prize-winning chemist and lunar scientist, expresses his attitude: "I do not know the origin of the moon. I'm not sure of my own or any other's models. I'd lay odds against any of the models proposed being correct."[33]

Baldwin ended a chapter on the moon's origin in 1965 with the following conclusion, and there has been no essential change in the picture since that time:

> We are thus left on the multipointed horns of a dilemma. There is no existing theory of the origin of the moon which gives a satisfactory explanation of the earth-moon system as we now have it. The moon is not an optical illusion or mirage. It exists and is associated with the earth. Before 4.5 billions years ago, the earth did not exist. Somehow, in this period of time, the two bodies were formed and became partners. But how?[34]

In considering the weaknesses of naturalistic lunar-origin hypotheses, it is well to realize that only a relatively few individuals originate and defend these broad theories. Although origin views receive first priority in the news media, the vast majority of scientists are not strongly involved in the debate over such theories. In fact, a subject such as lunar origins simply is not representative of scientific inquiry, since it is not reproducible. As one Apollo engineer explained it,

> You've got to realize that we've lived with some of these theories for so long that they don't mean much to us anymore. To a large extent these are matters of pure speculation. We've heard the same old people spin out the same old cobwebs and speculations for years without adding much to them.[35]

---

[32]"Another Origin for the Moon," *Science News*, 109:14 (April 3, 1976), 217.

[33]Robert Treash, "Magnetic Remanence in Lunar Rocks," *Pensée*, 2:2 (May, 1972), 22. Urey made the statement during a talk at the University of California at San Diego, as recorded by Robert Treash.

[34]Baldwin, *A Fundamental Survey of the Moon*, pp. 42-43.

[35]I. I. Mitroff, *The Subjective Side of Science: A Philosophical Inquiry into*

An interesting survey has been made of the scientists and engineers who were directly involved with the Apollo program. A cross-section of these experts were asked to rank the three main origin theories as to preference. Of the sample,

> 38 percent refused to select any lunar origin theory, expressing no interest in the subject.

> 62 percent complied by ranking the theories, though 20 percent of these mentioned little interest in origins.

> No single theory stood out as a majority choice with any statistical significance.[36]

The Apollo efforts certainly have provided a vast resource of scientific data, but a precise interpretation of origins is totally lacking. B. M. French, Program Chief for Extraterrestrial Materials Research, summarizes:

> In spite of everything that we have learned during the last few years, we still cannot decide between these three [lunar origin] theories. We will need more data and perhaps some new theories before the origin of the Moon is settled to everyone's satisfaction.[37]

Perhaps the reason for this uncertainty may be that *all* of the naturalistic origin theories are incorrect! The common assumption of these explanations is that the moon came into being by *accident*, in direct contradiction to the Genesis account.

When all other ideas are seen to fail, the world is faced with the full power and truth of the Genesis record of the creation of the moon and of the universe. It is the one proposition that fits all known facts. Unfortunately it is also the one proposition that is almost universally scorned and suppressed, in spite of the fact that those who do so are empty-handed.

---

the Psychology of the Apollo Moon Scientists (Amsterdam, Netherlands: Elsevier Scientific Publishing Co., 1974), p. 60. For a short summary of this survey, cf., I. I. Mitroff, "Studying the Lunar-Rock Scientists," *Saturday Review*, 2:4 (Nov. 2, 1974), 64-65.

[36] I. I. Mitroff, *The Subjective Side of Science*, p. 58.

[37] Berian M. French, *The New Moon: A Window on the Universe* (Washington, D.C.: NASA, 1975), p. 11.

CHAPTER III

The Genesis Record of the Moon's Creation

"And God made two great lights; the greater light to rule the day, and the lesser light to rule the night . . . ."
Genesis 1:16

"The revealed truths, if they be burdens to reason, are but such burdens as feathers are to a hawk, which, instead of hindering his flight by their weight, enable him to soar toward heaven, and take a larger prospect than, if he had no feathers, he could possibly do."

Robert Boyle (1627-1691)

The moon is referred to about 30 times in the Old Testament and 9 times in the New Testament.[1] In almost every instance, the theological significance of the moon for mankind is set forth. This should not come as a surprise to those who are familiar with the basic teachings of Scripture, for God's Word constitutes the divinely authoritative frame of reference within which all human activities, including science and engineering, are to be conducted, in accordance with

---

[1] In the Old Testament, the most common Hebrew word translated "moon" is *yā-rēaḥ* (very similar to the word for "moon" in other ancient Semitic languages). Three times the Hebrew word used is *lᵉḇānâ*, literally, "white one." In the New Testament the Greek word is *selēnē*. For a discussion of "month" and "new moon" see p. 148.

God's original mandate to Adam and Eve and their descendants to "subdue" the earth and to rule over it (Gen. 1:28).

Apart from such *direct, special revelation from God,* men
can never hope to discern the true significance of things or
the ultimate purpose of life, regardless of how much *indirect,
general revelation* is available. "For with thee is the fountain
of life" stated the psalmist, for "in thy light shall we see
light" (Ps. 36:9).

THE DOUBLE-REVELATION THEORY

Before the specific Biblical revelation concerning the creation of the moon is analyzed, it is essential to face a fundamental fallacy that is shared by many Christian theologians
and scientists with regard to the relation of *special* (or Biblical) revelation to *general* (or natural) revelation. For lack of a
better term, the authors have chosen to describe this error as
"the double-revelation theory." Briefly stated, proponents of
this theory maintain that God has given to man two distinct
and ultimate revelations of truth, each of which is fully
authoritative in its own realm: the revelation of God *in Scripture* (i.e., special revelation) and the revelation of God *in
nature* (i.e., general revelation). Although these two revelations differ greatly in their character and scope, they cannot
appear to intelligent men to contradict each other, since they
are given by the same self-consistent God of truth. *The theologian* is the God-appointed interpreter of Scripture, and *the
scientist* is the God-appointed interpreter of nature, each having specialized lenses for reading the true message of the
particular "book of revelation" which he has been called
upon to study.

Those who endorse the double-revelation theory also maintain that whenever there is an apparent conflict between the
conclusions of the scientist and the conclusions of the theologian, especially with regard to such questions as the origin
of the universe, the solar system, the moon, the earth, plant
life, animal life, and man, the theologian must rethink his
interpretation of Scripture at these points in such a way as to
bring the Bible into harmony with the scientists' consensus,
since the Bible is not a textbook of science and these problems overlap the territory in which science alone must give us

the detailed and authoritative answers.

It is held that such is necessarily the case, because if a grammatical/historical interpretation of the Biblical account of the creation of the moon, for example, should lead the Bible student to adopt conclusions that are contrary to the prevailing views of trained scientists concerning the origin and nature of the material universe, then he would be guilty of making God a deceiver of mankind in these vitally important matters.[2] But a God of truth cannot lie. Therefore, the Genesis creation account must be interpreted in such a way as to bring it into agreement with the generally accepted views of contemporary scientists. The early chapters of Genesis, we are told, were written primarily to give us answers only to such "spiritual" questions as "Who?" and "Why?" Scientists, however, must answer the important questions, "When?" and "How?"[3]

Limitations of the Scientific Method

Though this theory has gained considerable popularity in certain Christian circles, it has failed to come to grips with *major theological and scientific realities.* In the first place, proponents of the double-revelation theory fail to give due recognition to the tremendous limitations which inhibit the scientific method when applied to the study of ultimate origins. In the very nature of the case, the scientific method (which analyzes the processes of nature in observable and repeatable events) is incapable of coping with once-for-all and utterly unique events, or even the moral and spiritual (and thus empirically elusive) realities that give significance to human endeavor. It fails most conspicuously, however, when an attempt is made to employ it in analyzing the supernatural and miraculous acts of God (as recorded in Scripture) which form the foundation pieces of the Judeo-Christian world

---

[2]For a recent critique of this argument see Lewis H. Worrad, Jr., "God Does Not Deceive Men," *CRSQ*, 13:4 (March, 1977), 199-201. Cf. John C. Whitcomb, *The Early Earth* (Grand Rapids: Baker Book House, 1972), pp. 27-28; and John W. Klotz, *Modern Science in the Christian Life* (St. Louis: Concordia Publishing House, 1961), pp. 115-16.

[3]See Appendix III for a representative list of those whose writings reveal a basic commitment to the double-revelation theory.

view.

Those who exclusively employ the scientific method in historical sciences (e.g., paleontology) uncritically apply this method in a uniformitarian manner by extrapolating present natural processes forever into the past. Furthermore, they ignore the possible anti-theistic bias of the scientist himself as he handles the facts of nature in arriving at a *cosmology* (i.e., a theory concerning the basic structure and character of the universe) and a *cosmogony* (i.e., a theory concerning the origin of the universe in its parts). To the extent that such theorists fail to give careful and honest recognition to these essential limitations of the scientific method and of the investigator himself, they fail to give a true and undistorted picture of reality as a whole, and they fail also to point men to the only true source for understanding its mysteries.[4]

Scientific Obstacles to Cosmic Evolutionism

In the second place, advocates of the double-revelation

---

[4]Philosophers and logicians whose critiques of the Neo-Darwinian model of origins have gained wide recognition in recent years include: Thomas Bethell, "Darwin's Mistake," *Harper's Magazine* 252:1509 (February, 1976), 70-75; Lecomte du Noüy, *Human Destiny* (New York: Longmans, Green and Co., Inc., 1947); Marjorie Grene, *The Knower and the Known* (London: Faber & Faber, 1966), pp. 193-200; Gertrude Himmelfarb, *Darwin and the Darwinian Revolution* (New York: Doubleday and Co., Inc., 1962); Stanley L. Jaki, "The Role of Faith in Physics," *Zygon* 2:2 (June, 1967) 187-202, and *Science and Creation: From Eternal Cycles to An Oscillating Universe* (New York: Science History Publications, 1974); Robert E. Longacre, *An Anatomy of Speech Notions* (Lisse, Netherlands, Box 168: The Peter de Ritter Press, 1976); Norman Macbeth, *Darwin Retried* (Boston: Gambit, Inc., 1971 [see the analysis of Macbeth's arguments by William D. Stansfield, *The Science of Evolution*, New York: Macmillan Pub. Co., 1977, pp. 573-78]); Sir Karl Popper, *Conjectures and Refutations* (New York: Basic Books, Publishers, 1962); Richard Spilsbury, *Providence Lost: A Critique of Darwinism* (London: Oxford University Press, 1974); Anthony Standen, *Science is a Sacred Cow* (New York: E. P. Dutton and Co., Inc., 1950 [cf. Frank Trippett, "Science: No Longer A Sacred Cow," *Time* 109:10 (March 7, 1977), 72-73]). For a list of 12 additional important books and articles that deal with the logical status of evolutionary theory, see *Christianity Today* 21:18 (June 17, 1977), 15.

Christian theologians and philosophers whose writings have contributed significantly to the current renaissance of Biblical creationism include: Greg L. Bahnsen, "On Worshipping the Creature Rather Than the Creator," *The Journal of Christian Reconstruction* [P. O. Box 368, Woodland Hills, Calif. 91365; hereinafter referred to as *JCR*] ed. Gary North, 1:1 (Summer, 1974), 81-127; Louis Berkhof, *Systematic Theology* (Grand Rapids: Wm. B. Eerdmans Publishing Co., 1955), pp.

theory overlook the *insuperable scientific problems* which continue to plague currently popular naturalistic/evolutionary theories concerning the origin of the material universe and of living things. The impossibility of explaining mechanistically the biosphere that surrounds us at such close range on the planet earth continues to be an embarrassment to consistent evolutionists.

The thermodynamic and mathematical barriers to a chance transition from non-life to life in a primeval sea, the debilitating and even lethal effects of the vast majority of mutations, the large and as yet unbridged gaps between the various kinds of plants and animals in the fossil record, and the clear evidence of global catastrophes (rather than generally uniform processes) in the formation of coal seams and other fossil strata, have all contributed to the comparatively recent, widespread reappraisal of Lyellian/neo-Darwinian models of geology and paleontology.

Dissenting voices from within the empire of evolutionism

150-64; Gordon H. Clark, *The Philosophy of Science and Belief in God* (Nutley, N.J.: Presbyterian and Reformed Publishing Co., 1964); Charles A. Clough, *Laying the Foundation* (Lubbock, Texas: Lubbock Bible Church [3202 34th St., Lubbock, Texas 79410] 1973), "A Calm Appraisal of *The Genesis Flood*" (Unpublished Th.M. thesis, Dallas Theological Seminary, 1968), and "Biblical Presuppositions and Historical Geology: A Case Study," *JCR* 1:1 (Summer, 1974), 35-48; John J. Davis, *Paradise to Prison: Studies in Genesis* (Grand Rapids: Baker Book House, 1975); J. J. Duyvene de Wit, *A New Critique of the Transformist Principle in Evolutionary Biology* (Kampen, Netherlands: KoK, 1965); Weston Fields, *Unformed and Unfilled* (Nutley, N.J.: Presbyterian and Reformed Publishing Co., 1976); H. C. Leupold, *Exposition of Genesis* (Columbus, Ohio: The Wartburg Press, 1942); Harold Lindsell, "Where Did I Come From? A Question of Origins," *Christianity Today* 21:18 (June 17, 1977), 16-18; Byron Nelson, *After Its Kind* (Minneapolis: Bethany Fellowship, Inc., 1967); Vern S. Poythress, *Philosophy, Science and The Sovereignty of God* (Nutley, N.J.: The Presbyterian and Reformed Publishing Co., 1976); Robert L. Reymond, *A Christian View of Modern Science* (Nutley, N.J.: Presbyterian and Reformed Publishing Co., 1968); Alfred M. Rehwinkel, *The Flood* (St. Louis: Concordia Publishing House, 1951); Rousas J. Rushdoony, *The Mythology of Science* (Nutley, N.J.: The Craig Press, 1967); Francis A. Schaeffer, *Genesis in Space and Time* (Downers Grove, Ill.: InterVarsity Press, 1972); Gordon J. Spykman, "Biblical Authority and The Scientific Enterprise," *Pro Rege* 5:2 (December, 1976), 11-17; Cornelius Van Til, "The Doctrine of Creation and Christian Apologetics," *JCR* 1:1 (Summer, 1974), 69-80; John C. Whitcomb, *The Genesis Flood*, coauthored with Henry M. Morris (Nutley, N.J.: Presbyterian and Reformed Publishing Co., 1961); and Edward J. Young, *Studies in Genesis One* (Nutley, N.J.: Presbyterian and Reformed Publishing Co., 1964).

are being heard with increasing frequency;[5] but the most remarkable phenomenon of the past decade has been the veritable renaissance of Biblical creationism and catastrophism, with scores of Christian men of science contributing to this movement through their lectures and writings (see Appendix III for a representative listing). In the words of one opponent of this movement:

> Since the last century evangelical Christianity has been engaged in intense controversy over the relationship of geology and the Bible. This controversy received fresh impetus with the publication a decade ago of *The Genesis Flood* by John C. Whitcomb and Henry M. Morris. Their work has brought about a stunning

---

[5]See, for example, scientists such as Derek V. Ager, *The Nature of the Stratigraphic Record* (New York: John Wiley and Sons, 1973), who strongly opposes substantive uniformitarianism in paleontology, even at the risk that his arguments "should be read by some fundamentalist searching for straws to prop up his prejudices" (p. 19); Stephen Jay Gould, "This View of Life: The Problem of Perfection," *Natural History* 86:1 (Jan., 1977), 32-35, where it is stated that "organs of extreme perfection rank high in the arsenal of modern creationists" (p. 32); cf. "Evolution's Erratic Pace," *Natural History* 86:5 (May, 1977), 12-16, "The Return of the Hopeful Monster," *Natural History* 86:6 (June-July, 1977), 22-30; and "Punctuated Equilibria: The Tempo and Mode of Evolution Reconsidered," coauthored with Niles Eldredge, *Paleobiology* 3 (1977) 115-51; the distinguished French zoologist, Pierre P. Grassé, *l' Evolution du Vivant* (Paris: Editions Albin Michel, 1973); the late N. Heribert-Nilsson, *Synthetische Artbildung* (Lund, Sweden: Verlag CWE Gleerup, 1953 [an 1130-page, two-volume work in German, with a 100-page summarization in English available separately with an introduction by George F. Howe, Ph.D., c/o W. D. Burrowes, Box 5083, Stn. B, Victoria, B.C., Canada, V8R 6N3]); Allan O. Kelly and Frank Dachille, *Target: Earth* (Carlsbad, Calif.: Target Earth, 1953, cf. Robert S. Deitz, "Astroblemes," *Scientific American* 205:2 [August, 1961], 50-58); G. A. Kerkut, *Implications of Evolution* (New York: Pergamon Press, 1960), who insists that "much of the evolution of the major groups of animals has to be taken on trust" (p. 154); Johan B. Kloosterman, editor of *Catastrophist Geology* (Caixa Postal 41.003, Rio de Janeiro, Brazil), a journal dedicated to the study of discontinuities in earth history; Claude Levi-Strauss, *Structural Anthropology* and other volumes (cf. analysis in *Time* 89:26 [June 30, 1967], 34-35); Paul S. Moorhead and Martin M. Kaplan (eds.), *Mathematical Challenges to the Neo-Darwinian Interpretation of Evolution* (Philadelphia: Wistar Institute Press, 1967); Norman D. Newell, "Crises in the History of Life," *Scientific American* 208:2 (February, 1963), 77-92; P. H. Schindewolf, *Grundfragen der Paläontologie* (Stuttgart, 1950) and "Neokatastrophismus," *Deutsche Geol. Gesell. Zeitschr.*, Vol. 114, 430-45; and William R. Thompson, Introduction to *Charles Darwin: The Origin of Species*, Everyman's Library No. 811 (London: J. M. Dent and Sons, Ltd., and New York: E. P. Dutton and Co., Inc., 1958), pp. vii-xxv, reprinted in *Journal of the American Scientific Affiliation* [hereinafter referred to as *JASA*] 12:1 (March, 1960), 2-9, who asserted that "the success of Darwinism was accompanied by a decline in scientific integrity" (p. 8).

renaissance of flood geology in Christian circles. Their influence has led to the formation of such organizations as the Creation Research Society, the Creation Science Research Center, and the Institute for Creation Research. Journals, radio programs, lecture tours, and curriculum materials devoted to the propagation of flood geology and special creationism have proliferated and met with astounding success among ministers, educators, scientists, and laymen from a wide spectrum of denominations. The impact of this movement has been felt in the recent California education controversy.[6]

One notable challenge to the concept of a vastly ancient, slowly changing earth/moon system has been the discovery that the earth's powerful electro-magnetic field is rapidly decreasing. Accurate measurements of the absolute value or strength of the earth's magnetic field (its magnetic moment) have been made in various parts of the world continuously since 1835. The mathematical form which best fits the data is an exponential decrease with time (see Table III-1). This is the same kind of variation as the decay of a radioactive isotope and, like it, may be characterized by a half-life. The half-life of the earth's magnetic moment (the overall representation of the magnetic strength) is only about 1,400 years. Thus, assuming an exponential extrapolation for the sake of argument, 10,000 years ago the earth's magnetic field would have been 142 times as great as it is today—nearly equivalent to that of some magnetic stars! Of course, the earth could not have had such a large magnetic field without a huge sustaining nuclear energy source such as these magnetic stars have.

Thomas G. Barnes, Professor of Physics at the University of Texas in El Paso, the author of a college textbook on electricity and magnetism,[7] points out that

the only alternative to a "young age" for the earth's magnetic field is to deny the existence of decay in the earth's magnetic field, not a very astute stand for a scientist in view of the strong

---

[6]Davis A. Young, *Creation and the Flood: An Alternative to Flood Geology and Theistic Evolution* (Grand Rapids: Baker Book House, 1977), p. 7.

[7]Thomas G. Barnes, *Foundations of Electricity and Magnetism* 2nd ed. (Boston: D. C. Heath and Company, 1965); third edition, enlarged, available from the author (2115 N. Kansas St., El Paso, Texas 79902).

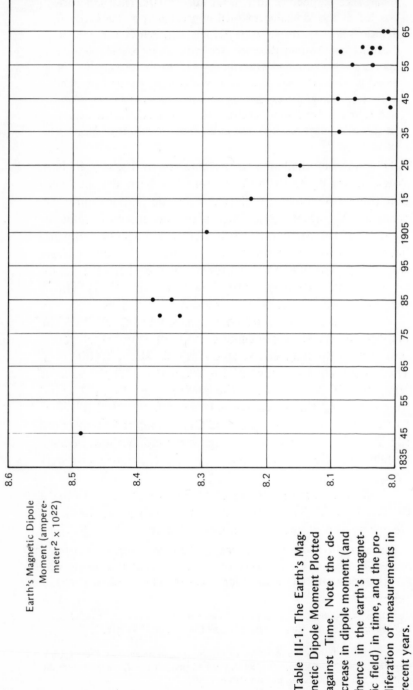

Earth's Magnetic Dipole Moment (ampere-meter^2 × 10^{22})

Year

Table III-1. The Earth's Magnetic Dipole Moment Plotted against Time. Note the decrease in dipole moment (and hence in the earth's magnetic field) in time, and the proliferation of measurements in recent years.

physics upon which Sir Horace Lamb's theory is based and the 130 years of worldwide real-time data to substantiate it.[8]

An age for the earth of only 10,000 years seems incredibly short for uniformitarian scientists who are accustomed to think in terms of billions of years. However, as Barnes explains,

> if one goes back 21,000 years, the power dissipation (which is proportional to the square of the field strength and associated current strength) would have been more than *a billion times* its present value of 813 megawatts. That would be quite a lot of heat buildup in the core of the earth and would perhaps have fractured the earth. Certainly if one goes much farther back with that exponential increase there is no doubt that the earth would have blown apart.[9]

Professor Barnes also claims that the notion that the earth's magnetic field reverses itself is not supported by geological evidence or physical theory. From a geological perspective, he has suggested a number of ways in which rocks can have their own magnetic orientation induced without help from the earth's magnetic field. From the standpoint of physical theory, he questions whether the earth's electromagnet ever could have reversed itself.

This discovery provides enormous leverage against purely naturalistic theories of the origin of the earth and its living inhabitants, and serves to accentuate the tragedy of such positions of compromise as the double-revelation theory. A clear recognition of the evidence from special revelation in the Bible concerning the comparatively recent creation of the earth would have eliminated the deep-seated resistance to such discoveries which characterizes so much of the theologi-

---

[8]Thomas G. Barnes, *Origin and Destiny of the Earth's Magnetic Field* (San Diego: Institute for Creation Research [2716 Madison Avenue, San Diego, Calif. 92116]), p. 25. For further discussion of the earth's decreasing magnetic field, see K. L. McDonald and R. H. Gunst, *Earth's Magnetic Field 1935 to 1965* (ESSA Technical Report, IER 46-IES, 1967, available from Superintendent of Documents, U. S. Government Printing Office, Washington, D.C. 20402). See also "A New Doomsday?" *Time* 91:11 (March 15, 1968), 36,38; and Henry M. Morris, *Scientific Creationism* (San Diego: Creation-Life Publishers, 1975), pp.. 157-58.

[9]Personal communications from Dr. Thomas G. Barnes, Department of Physics, The University of Texas at El Paso, May 23 and October 6, 1977.

cal and scientific establishment today.[10]   The implications of all this for the recent origin of the moon are momentous, for the theory of an ancient moon could hardly survive the discovery that the earth itself is recent in origin.[11]

A Fatal Neglect of Biblical Revelation

In the third place, and most serious of all from the Biblical perspective, most proponents of the double-revelation theory underestimate and even deny the supreme authority and self-evident clarity of God's special revelation in Scripture. The Biblical record of physical and biological origins is thoroughly historical and amazingly detailed. It may come as a great surprise even to many Christians to discover that in the first eleven chapters of Genesis

> there are 64 geographical terms, 88 personal names, 48 generic names and at least 21 identifiable cultural items (such as gold, bdellium, onyx, brass, iron, gopher wood, bitumen, mortar, brick, stone, harp, pipe, cities, towers). . . . The significance of this list may be seen by comparing it, for example, with "the paucity of references in the Koran. The single tenth chapter of Genesis has five times more geographical data of importance than the whole of the Koran." Every one of these items presents us with the possibility of establishing the reliability of our author. The content runs head-on into a description of the real world rather than recounting events belonging to another world or level of reality.[12]

Agreeing with a detailed and profound analysis of the opening chapter of the Bible by Edward J. Young,[13] Dr. Kaiser

---

[10]See below, note 48.

[11]The consensus among astronomers today is that the earth and moon originated at approximately the same time. See pp. 46, 92.

[12]Walter G. Kaiser, Jr., "The Literary Form of Genesis 1-11," in *New Perspectives on the Old Testament,* ed. by J. Barton Payne (Waco, Texas, Word, Inc., 1970), 59. For preliminary analyses of the spectacular archeological discovery of ancient Ebla in North Syria, and its significance for the historicity of Genesis, cf. Giovanni Pettinato, "The Royal Archives of Tell Mardikh-Ebla," *Biblical Archeologist* 39:2 (May, 1976), 44-52; and Paolo Matthiae, "Ebla in the Late Early Syrian Period: The Royal Palace and The State Archives," *Biblical Archeologist* 39:3 (September, 1976), 94-113.

[13]Edward J. Young, *Studies in Genesis One* (Nutley, N.J.: Presbyterian and Reformed Publishing Co., 1964), p. 105.

concludes:

> The decision is easy: Genesis 1-11 is prose and not poetry. The use of the *waw* consecutive with the verb to describe sequential acts, the frequent use of the direct object sign and the so-called relative pronoun, the stress on definitions, and the spreading out of these events in a sequential order indicates that *we are in prose and not in poetry*. Say what *we* will, the author plainly intends to be doing the same thing in these chapters that he is doing in chapters 12-50. If we want a sample of what the author's poetry, with its Hebrew parallelism and fixed pairs, would look like, Genesis 4:23-24 will serve as an illustration.[14]

Thus, Genesis 1-11 is detailed, accurate, authoritative, *non-poetic* history. For that matter, even *poetic* history in the Bible (such as Psalms 78, 105, 106, and 136, which narrate events from Israel's past) intends to be taken literally, with normal allowance for figures of speech, and is a fully authoritative expression of the mind of God. Therefore, pre-Abrahamic Biblical history stands firmly upon the rock of objectivity and cannot be twisted to suit the whim of the interpreter. By making so many detailed statements about history, chronology and geography (as well as astronomy, geology and zoology), the Author of Genesis is showing His hand as One *who expects to be taken seriously and who actually provides material suitable for investigation* (in contrast to the sacred writings of other ancient religions).

The enormous significance of these facts is only now beginning to dawn upon philosophers of science. Neo-Darwinists have been increasingly criticized for such tautologies as the "survival of the fittest," for the ubiquity of their secondary assumptions and modifications to avoid the negative impact of the second law of thermodynamics and other aspects of reality, and for the suspicious flexibility of a scheme that can come up with *ad hoc* explanations of exactly opposite phenomena (e.g., the long neck of the giraffe and the short neck of the hippopotamus). It is the very vagueness and extreme adaptability of evolutionism that qualifies it for Sir Karl Popper's famous definition of a "bad theory":

> A theory which is not refutable by any conceivable event is non-

---

[14]W. C. Kaiser, "The Literary Form of Genesis 1-11," 59-60.

scientific. Irrefutability is not a virtue of a theory (as people often think) but a vice. Every "good" scientific theory is a prohibition; it forbids certain things to happen. The more a theory forbids, the better it is. Confirmations should count only if they are the result of *risky* predictions; that is to say, if, unenlightened by the theory in question, we should have expected an event which was incompatible with the theory—an event which would have refuted the theory.[15]

Biblical creationism is, strictly speaking, *non-scientific* (e.g., Heb. 11:3—"*Through faith* we understand that the worlds were framed by the word of God . . . ."). *But so is evolutionism!* Neither model can be *empirically proved*, for each one deals with non-observable ultimate origins, and flows logically from a philosophico-religious world view which is a faith proposition: theism or non-theism. The vast superiority of the creation model, however, lies in the astounding confirmations of its "risky prediction," to use Sir Karl Popper's expression. Firmly locked into the Creation/ Curse/Flood model of Genesis are such predictions as the impossibility of cross-breeding or developing different "kinds" of living things, the general disintegration of energy systems throughout the universe,[16] and the post-Adamic formation of fossil strata through the hydrodynamic forces of a global deluge.[17] These detailed and self-consistent Biblical

---

[15]Karl R. Popper, *Conjectures and Refutations* (New York: Basic Books, Publishers, 1962), p. 36. Cf. Norman Macbeth, *Darwin Retried*, p. 99.

[16]For helpful presentations of the full implications of the Second Law of Thermodynamics for cosmogony, cf. A. E. Wilder-Smith, *Man's Origin, Man's Destiny* (Wheaton, Ill.: Harold Shaw Publishers, 1968), pp. 55-109, and *The Creation of Life* (Wheaton, Ill.: Harold Shaw Publishers, 1970), pp. 56-57, 103-17; David Penny, "The Implications of the Two Laws of Thermodynamics in the Origin and Destiny of the Universe," *CRSQ* 8:4 (March, 1972), 261-69; George Mulfinger, Jr., "Review of Creationist Astronomy," *CRSQ* 10:3 (December, 1973), 170-75; Emmett L. Williams, "Thermodynamics: A Tool for Creationists," *CRSQ* 10:1 (June, 1973), 38-44; Robert L. St. Peter, "Let's Deflate the Big Bang Hypothesis," *CRSQ* 11:3 (December, 1974), 143-55; H. L. Armstrong, "Use of the Second Law of Thermodynamics in Macroscopic Form in Creation Studies," *CRSQ* 12:2 (September, 1975), 103-06; George Mulfinger, "Critique of Stellar Evolution," in *Speak to the Earth*, ed. by George F. Howe (Nutley, N.J.: Presbyterian and Reformed Publishing Co., 1975), pp. 409-46 (reprinted from *CRSQ* 6:5 [June, 1970]); and Henry M. Morris, *Scientific Creationism*, pp. 37-46.

[17]Cf. John C. Whitcomb and Henry M. Morris, *The Genesis Flood* (Nutley, N.J.: Presbyterian and Reformed Publishing Co., 1961).

Near-vertical photograph of cratered lunar surface near the terminator at 155° W and 3° S, far side. (NASA)

The terraced crater Langrenus from an altitude of 150 miles. The crater is about 85 miles in diameter. Note the sharp circular craters in the area. (NASA)

Astronaut Aldrin (Apollo 11) leaves the Apollo 11 Lunar Module. (NASA)

The dusty surface at Tranquillity Base recorded footprints similar to damp sand. Although superficially soft, the material strongly resisted penetration by coring tubes. (NASA)

Astronaut Irwin (Apollo 15) saluting beside flag, Lunar Module in center, Rover on right. Mount Hadley in the background rises 4.6 kilometers above the plain. (NASA)

The Apollo 15 crew saw this thin crescent earth rising above the lunar horizon. At the same time, we on earth were seeing a nearly full moon. (NASA)

Astronaut Aldrin walks near a gold foil-wrapped footpad of the Lunar Module Eagle. Note the footprints in the foreground. (NASA)

Astronaut Irwin with Rover, Mount Hadley in the background. (NASA)

concepts are utterly unique in the history of thought, and stand unexpectedly and sharply against the background of ancient Near Eastern cosmologies. Even more to the point, they have been repeatedly confirmed by observation and experimentation. Recently, Morris[18] and Bliss[19] have provided helpful analyses of the basic predictions and comparative value of these two models.

Biblically informed Christians, therefore, have more reasons than ever before to look with profound dismay at the double-revelation theory and other attempted "harmonizations" of Genesis and evolutionism in its various forms. Scientists who desire to find answers to the questions of ultimate origins, meaning, and destiny cannot succeed by positioning themselves in opposition to God's infallible Word. Nor can theologians! Therefore, scientists and theologians who seek for truth in these vitally important areas of research must function within the universally valid guidelines of spiritual and natural laws. Scripture and science are *not* mutually contradictory; they are *not* in competition with each other; they have been designed by God in such a way that the latter is defined, conditioned, and guided by the former, resulting in genuine scientific truth. The harmony cannot be between Genesis and evolutionism; the harmony is rather between Genesis and objective, empirical science. It is in this realm of reality that the supposed warfare between science and theology comes to an end.[20]

Consequently, Christians must abandon all hope of formulating a scientifically valid cosmogony if they fall prey to the popular notion that science provides an *independent* and *equally authoritative* source of information with the Bible concerning the creation of the universe, the solar system, and the earth with its living forms; and that science alone is competent to tell us *when* and *how* such things occurred, while

---

[18] Henry M. Morris, *Scientific Creationism* (San Diego: Creation-Life Publishers [P. O. Box 15666, San Diego, Calif. 92115], 1974), pp. 1-16.

[19] Richard B. Bliss, *Origins: Two Models* (San Diego, Calif. 92115: Creation-Life Publishers, 1976).

[20] Cf. Gordon J. Spykman, "Biblical Authority and the Scientific Enterprise," *Pro Rege* 5:2 (December, 1976), 11-17.

the Bible merely informs us in "spiritual" terms (or, as neo-orthodox theologians would express it, "suprahistorical" language) as to *who* brought the universe into existence and *why*.

The truth of the matter is that the Word of God not only provides us with the only reliable source of information as to the *who* and the *why* of these great events, but also provides essential information concerning the *when* and the *how*.[21] Such an assertion has, of course, been widely resisted in modern times, usually with ominous references to Luther's rejection of Copernicus[22] and the Pope's persecution of Galileo.[23] But in these famous (and thankfully rare) examples of opposition to genuine scientific discovery in the name of Scripture, a more careful examination of the facts reveals that in these and similar cases the Bible's own guidelines to its interpretation had been neglected.

From Genesis to Revelation the Bible consistently avoids (for the sake of effective communication to mankind) highly technical teaching of scientific data or concepts. Nevertheless—and this is crucial to the entire discussion—*the Bible provides perfectly accurate descriptions of things by the use of the language of appearance.* This has been frequently chal-

---

[21] For the *when* of creation, see note 48 below. For the *how* of creation, see pp. 70-83.

[22] Luther never officially opposed the views of Copernicus. In fact, two friends and disciples of Copernicus were on the faculty of Luther's university at Wittenberg. See R. Hooykaas, *Religion and the Rise of Modern Science*, p. 122; John W. Klotz, *Modern Science in the Christian Life* (St. Louis: Concordia Publishing House, 1961), pp. 86-89; and R. Laird Harris, "Copernicus and the Church," *Christianity Today* 17:24 (Sept. 14, 1973), 4-5.

[23] The entire debate between Galileo and the Church as to whether or not the earth moves around the sun was Biblically irrelevant. It is true that the psalmist stated that "the earth shall not be moved" (Psa. 93:1 and 96:10). However, the psalmist, speaking of himself, also asserted: "I shall not be moved" (Psa. 10:6, 16:8, 30:6, 55:22, 62:2, 66:9, and 121:3). The obvious point is that he did not anticipate being swerved from the path God had marked out for him. Likewise, the earth cannot be diverted from its God-ordained functions, such as its orbital movement around the sun. For further analysis of the Galileo controversy, see Charles E. Hummel, "The Scientific Revolution of the 16th and 17th Centuries," *JASA* 20:4 (Dec., 1968), 101; Jerome J. Langford, *Galileo, Science and the Church*, rev. ed. (Ann Arbor: The University of Michigan Press, 1971), pp. 1-158; and R. Hooykaas, *Religion and the Rise of Modern Science*, pp. 124-26.

lenged, but never successfully.

A perfect example of this principle is the account of the creation of the sun and moon in Genesis 1:16. "And God made two great lights; the greater light to rule the day, and the lesser light to rule the night: he made the stars also." On an absolute scale, of course, the sun and moon are not "great lights" compared to many of the giant stars. In fact, the moon is not a "light" at all in the sense that the sun is a light. But from the perspective of earth-dwellers, the statement is vastly more meaningful than a technical astronomical analysis. Furthermore, the statement is perfectly accurate. There are only two *great* lights visible to the unaided eye, not three or ten.

Another outstanding example of this principle is found in Revelation 7:1, which speaks of the earth having "four corners." This does not suggest that the Bible subscribes to the flat-earth concept, for the following phrase explains that "the four corners" refer to "the four winds of the earth." Even today meteorologists use the four directions of the compass to describe wind movements, without thereby implying that the earth is flat.

Finally, and perhaps most famous of all, are Biblical statements that refer to "the rising of the sun" (e.g., Rev. 7:2 NASB, cf. Ps. 19:4-6). Does this mean that the Scriptures teach geocentrism? Not necessarily, for this is a language of appearance so appropriate that it cannot be improved upon even by astronomers of our day.[24]

The Scriptures explain why the unaided human intellect (however brilliant) is utterly incompetent to arrive independently at the correct answers to such stupendously important questions as *ultimate origins.*[25] God's challenge to Job still echoes down the centuries: "Where wast thou when I laid the

---

[24]For a helpful study of this crucial issue, see R. Laird Harris, "The Bible and Cosmology," *Bulletin of the Evangelical Theological Society* 5:1 (March, 1962), 11-17; and a letter to the editor of *JASA* 21:3 (September, 1969), 92-93. For a solution to the supposed mathematical error in 1 Kings 7:23 (cf. 2 Chron. 4:2), see Henry M. Morris, *Many Infallible Proofs* (San Diego: Creation-Life Publishers, 1975), pp. 243-44.

[25]Cf. Romans 1:18-23; 3:11; 1 Corinthians 1:19-29; 2:14; Hebrews 11:1-6; 2 Peter 3:3-5.

foundations of the earth? declare, if thou hast understanding" (Job 38:4). "Lift up your eyes on high, and behold who
hath created these things, that bringeth out their host by
number . . . . there is no searching of his understanding" (Isa.
40:26-28). Christ's condemnation of the pseudo-theologians
of His day adequately expresses the fundamental problem we
all confront within ourselves, whatever our professional experience or reputation might be: "You are mistaken, not
understanding the Scriptures, or the power of God" (Matt.
22:29, NASB).

No one appreciates God's general revelation of Himself to
mankind in the material universe more than the Biblically
trained Christian. Loyalty to God's special revelation in the
Bible is in no way embarrassed or compromised by the recognition that God is allowing men (through the use of God-
given attributes and abilities) to discover important things
about Himself through nature. In fact, the Bible itself em-
phasizes this wonderful fact. "The heavens declare the glory
of God; and the firmament sheweth his handywork. Day
unto day uttereth speech, and night unto night sheweth
knowledge" (Ps. 19:1-2). More ominous, however, is the fact
that God holds men responsible for hindering and opposing
the marvelous things He is telling them about Himself in
nature:

> For the wrath of God is revealed from heaven against all ungodli-
> ness and unrighteousness of men, who hold (Greek: *katechontōn*,
> hold back, hinder, oppose) the truth in unrighteousness; because
> that which may be known of God is manifest in them; for God
> hath shewed it unto them. For the invisible things of him from
> the creation of the world are clearly seen, being understood by
> the things that are made, even his eternal power and Godhead; so
> that they are without excuse (Rom. 1:18-20; cf. v. 25).

While acknowledging the tremendous significance of God's
revelation of Himself to mankind through the testimony of
the natural universe, we must at the same time insist that
there are a great number of supremely important truths that
the material universe can never reveal to the searching eye of
man, even if he could bring an unfallen mind and a pure heart
to the investigation of its wonders. It is for this very reason
that God, in His infinite grace a   love, has given to mankind

His written Word, the final and only detailed revelation of His attributes, His great redemptive plan, His purposes for the universe and its various parts, and other vitally important truths that are completely outside the scope of empirical scientific investigation (cf. 1 Cor. 2:9-10). In other words, cosmogony, cosmology, and metaphysics, in the ultimate sense of these terms, must grope in darkness apart from God's special revelation in Scripture. The true scientist, therefore, no less than the true theologian, must confess with the psalmist: "Thy word is a lamp unto my feet, and a light unto my path" (Ps. 119:105; cf. Prov. 1:7, 2 Cor. 10:5, Col. 2:3).

In view of all this, the Christian may have perfect confidence that all genuine scientific discoveries will be in perfect accord with the detailed, clear and obvious teachings of God's written Word. Some, indeed, will consider this to be an unwarranted restriction on their intellectual freedom, and a stumbling block in their pathway as they seek to "follow truth wherever it may lead." But exactly the opposite results will be experienced by those who allow the Scriptures to be their guide in such matters, for the omniscient, truthful, and totally dependable Creator and Lord of mankind has promised us: "And ye shall know the truth, and the truth shall make you free" (John 8:32).

THE CREATION OF THE MOON

How, then, can a Christian theologian and/or scientist, standing upon solidly established principles of theology, metaphysics and epistemology, develop a detailed system of lunar science that takes all the relevant facts into consideration? Since all facts must ultimately be God-oriented facts,[26] the moon, like all other facts of human experience,

---

[26]"The argument between Christians and non-Christians involves every fact or it does not involve any fact. If one fact can be interpreted correctly on the assumption of human autonomy then all facts can. If the Christian is to be able to show the non-Christian objectively that Christianity is true and that those who reject it do so because they hold to that which is false, this must be done everywhere [including lunar studies—the present authors] or else it is not really done anywhere." Cornelius Van Til, *The Defense of the Faith* (Nutley, N.J.: Presbyterian and Reformed Publishing Co., 1955), p. 191; cf. pp. 58-63. Cf. also, Gary North, ed., *Foundations of Christian Scholarship: Essays in the Van Til Perspective* (Vallecito, Calif.: Ross House Books [P. O. Box 67], 1976).

must first be seen within the context of special divine revelation. Only when this indispensable frame of interpretive reference has been thoroughly explored can the investigator hope to find dependable and truly significant answers with respect to the various details.

Obviously God knew that men, given enough time, would invent the machinery and develop the skills necessary for examining the structure and substance of the moon. But God also knew that only His revelation in Scripture could supply such absolutely essential ingredients in lunar science as the origin, the true significance and purpose, and the final destiny of the earth's great satellite.

Biblical revelation concerning the moon centers on three basic concepts: (1) the manifestation of God's wisdom and power through the direct creation of the moon; (2) the dependability and destiny of the moon as a divinely ordained sign; and (3) the spiritual disaster of moon worship. The second and third concepts will be discussed in Appendix I and Appendix II, but the first of these aspects of God's revelation in Scripture concerning the lesser light that rules the night (Gen. 1:16) must now be considered.

The Manifestation of God's Wisdom and Power Through the Direct Creation of the Moon

Biblical writers freely expressed their fascination with the spectacular beauty of the moon. Four thousand years ago Job referred to "the moon walking in brightness" across the night sky (Job 31:26). A thousand years later Solomon described his lovely Shulamite bride-to-be as "fair as the moon" (S. of Sol. 6:10). A thousand years still later the Apostle Paul observed that "there is one glory of the sun, and another glory of the moon" (1 Cor. 15:41), thus agreeing with the psalmist that "the heavens declare the glory of God" (Ps. 19:1).

As a young lad tending his father's flocks by night, David had numerous opportunities to scan the night skies from the hills around Bethlehem. His mind was conditioned by Biblical rather than pagan concepts of the origin and significance of the heavens; therefore, what he saw in the clear sky above simply overwhelmed him. "When I consider thy heavens, the work of thy fingers, the moon and the stars, which thou hast ordained; what is man, that thou art mindful of him? and the son of man, that thou visitest him? (Ps. 8:3-4).

In one of the greatest of the nature psalms, our magnificent lunar neighbor is not by-passed in the catalog of universal wonders.

> He appointed the moon for seasons . . . Thou makest darkness, and it is night . . . O Lord, how manifold are thy works! in wisdom hast thou made them all . . . the glory of the Lord shall endure for ever: the Lord shall rejoice in his works (Ps. 104:19-20, 24, 31).

In fact, one of the truly great evidences that "his mercy endureth for ever" (Ps. 136:8) is that "the moon and stars . . . rule by night" (Ps. 136:9). It is because God is glorified in all His works of creation that the psalmist cries out: "Praise ye him, sun and moon: praise him, all ye stars of light" (Ps. 148:3).[27]

To the Hebrew mind, the concept of the creation of the moon (together with the sun and stars) served to magnify the glory of God because of *the special way* in which the opening chapter of the Hebrew Bible sets forth this cosmic event. In total contrast to the wide spectrum of creation concepts that characterizes modern Christendom, including such views as theistic evolutionism (with God providentially directing the cosmos toward higher levels of complexity throughout endless ages) and progressive creationism (with God occasionally introducing new kinds of plants and animals and finally man into a geologic timetable spanning billions of years) and the gap theory of Genesis 1:1-2 (with God creating a perfect world perhaps billions of years ago, destroying it all at the fall of Satan, and then re-creating it in six literal days),[28] the traditional Hebrew/Christian understanding of the opening

---

[27]"Scientific work that does *not* result in such praise is itself 'religious,' claiming in effect that the God of the Bible is *not* the cause of the phenomena (metaphysical claim), that God is *not* the cause of knowledge (epistemological claim), and that God *ought not* to be praised (ethical claim)." Personal communication from Charles A. Clough, 3202 34th St., Lubbock, Texas 79410, May 11, 1977. Cf. also Vern Poythress, "A Biblical View of Mathematics," in *Foundations of Christian Scholarship*, ed. Gary North, pp. 165-68.

[28]For a definitive critique of the Gap Theory of Genesis 1:1-2, as well as of similar views (e.g., gaps between the days of creation, a gap before Genesis 1:1,

chapter of Genesis has been simple and straightforward.

There is a very good reason for this: all other historical narratives in the Bible were understood in a normal manner (technically known as the historical/grammatical method of hermeneutics, which takes into full account not only the context of each passage but also all the known literary figures of speech),[29] and since there is no evidence of poetry in the first chapter of Genesis,[30] it seems rather obvious that God intended the chapter to be understood normally. If, on the other hand, we abandon the God-honored and time-honored method of historical/grammatical interpretation, then all hope of definitively determining what the opening statements of the Bible mean must be abandoned.

Now it is essential to recognize that the reason Biblical writers praised God for His work of creating the sun, moon, and stars is that they *did* understand the Genesis account! At least two things can be clearly discerned in the creation record that unveil the absolute glory of the Creator. First, *the astronomical bodies were created suddenly,* thus establishing the overwhelming uniqueness, in fact, the absolute ultimacy of God's power (omnipotence). Secondly, *the astronomical bodies were created after the earth and plant life had been created,*[31] thereby eliminating all potential competition (in

---

etc.), cf. Weston W. Fields, *Unformed and Unfilled* (Nutley, N.J.: Presbyterian and Reformed Publishing Co., 1976); and John J. Davis, *Paradise to Prison: Studies in Genesis,* pp. 37-57. See also J. C. Whitcomb, *The Early Earth,* pp. 115-34; and Henry M. Morris, *Scientific Creationism,* pp. 220-55.

[29] For a helpful analysis of the Biblical use of figures of speech, cf. Paul Lee Tan, *The Interpretation of Prophecy* (Winona Lake, Ind.: BMH Books, Inc., 1974), pp. 29-39; and the classic work by E. W. Bullinger, *Figures of Speech Used in the Bible* (Grand Rapids: Baker Book House, reprinted 1968).

[30] See the discussion, p. 63.

[31] It has been claimed frequently that the sun and moon were not created on the fourth day because the Hebrew verb used in Genesis 1:16 is 'āśāh ("made") rather than bārā' ("created") as in Genesis 1:1. However, this is a serious exegetical blunder. In a creation context, the two verbs are used synonymously, as any concordance will show. For example, marine creatures were "created" (v. 21) while land animals were "made" (v. 25). Surely this cannot mean that land animals were not created! Furthermore, these two verbs are used alternately to describe the very same events: Gen. 1:26 ("make") and 1:27 ("created"); Gen. 2:4a ("created") and 2:4b ("made"); Gen. 1:1 ("created") and Exod. 20:11

terms of a solar or lunar deity or even the modern secular "god" of cosmic evolutionism) for the claim of final sovereignty and deity.[32] The first of these two profoundly important facts of special revelation must now be analyzed in more detail.

The Instantaneous Creation of the Moon

The creation of the astronomical universe was not only *ex nihilo* (i.e., from no previously existing matter, as stated in Heb. 11:3),[33] but it was also, by the very nature of the case, *instantaneous.* Its origin could not, therefore, have been spontaneous or self-acting. The evolutionary concept of a gradual buildup of heavier and heavier elements throughout billions of years is clearly excluded by the pronouncements of Scripture.

In the first place, the immediate effect of God's creative word is emphatically stated in Psalm 33: 6,9—"By the word of the Lord were the heavens made; and all the host of them by the breath of his mouth. . . . For he spake, and it was done; he commanded, and it stood fast." There is certainly no thought here of gradual development, or trial and error process, or age-long, step-by-step fulfillment. In fact, it is quite impossible to imagine any time interval in the transition from absolute nonexistence to existence! Similarly: "And God said, Let there be light: and there was light" (Gen. 1:3). At one moment, there was no light anywhere in the universe;

("made"); Gen. 1:16 ("made") and Ps. 148:3,5 with Isa. 40:26 ("created"). Bruce K. Waltke concludes: "Moreover, it is clear that *'asah* and the other verbs may designate creation by fiat *ex nihilo.* The doctrine of *creatio ex nihilo* does not depend on the verb *bara'.* . . .The sun, moon and stars came into existence at the sole bidding of their Creator" ("The Creation Account in Genesis 1:1-3," *Bibliotheca Sacra* 132:528 [October, 1975], 337). Thus, it is exegetically impossible to date the moon's origin before the fourth day of creation week. Cf. Weston W. Fields, *Unformed and Unfilled,* pp. 53-74; and J. C. Whitcomb, *The Early Earth,* pp. 127-29.

[32]For further development of this crucial theological point, see Appendix II.

[33]Hebrews 11:3 certainly cannot mean that the physical substances that compose our visible universe consist of "invisible" atomic particles! Spiritual faith is certainly not required to accept the atomic theory of matter. The point of this key verse on creationism is that visible material substances did not exist in any form whatsoever, other than in the mind of an eternal and omniscient God, until He spoke the creative Word. Cf. Romans 4:17.

the next moment, there was! So spectacular is this creation event that the New Testament compares it to the suddenness and supernaturalness of conversion (2 Cor. 4:4-6; cf. 5:17). It may be confidently asserted that the idea of *sudden appearance* dominates the entire creation account (cf. Gen. 1:1,3,12,16,21,25,27; 2:7,19,22).

One is confronted, however, with numerous contemporary denials of this concept. It is frequently claimed, for example, that God's command to the earth to "bring forth grass, the herb yielding seed, and the fruit tree yielding fruit after his kind" (Gen. 1:11) implies a long process under the providential direction of God. But Numbers 17:8 is the true analogy to Genesis 1:11, for in one night Aaron's rod "budded, and brought forth buds, and bloomed blossoms, and yielded almonds" (cf. Jonah 4:6-10). It is significant that Russell W. Maatman, a proponent of the day-age theory, admits that "there is no doubt that *each creation event* was instantaneous. One moment a certain thing existed; the previous moment, it did not exist."[34] It must be made clear that theologians who emphasize the suddenness of God's *creative* acts and sign miracles do not thereby minimize the glory of God's non-miraculous *providential* works in human history (cf. Dan. 4:17 and the Book of Esther). Nevertheless, miracle and providence are *not* identical and dare not be confused. Thus, the conception of Jesus was sudden and supernatural while His birth was the result of a gradual and natural process carried out under the providential control of God. This distinction is profoundly important, for if the conception of Jesus is understood to be providential but not miraculous, the incarnation is denied and Christianity is destroyed (cf. 1 John 4:3; 2 John 7). Likewise, if the events of Genesis 1-2 are understood to be providential but not miraculous, creationism is not simply modified; it is destroyed.

This leads us to a second important consideration pertaining to the sudden creation of the astronomical universe, namely, the analogy of God's creative works in the person of Christ during His earthly ministry nearly two thousand years

---

[34] *The Bible, Natural Science and Evolution* (Grand Rapids: Baker Book House, 1970), p. 95.

ago in Palestine. Since the New Testament makes it clear that the universe was created through Christ, the Son of God (John 1:3,10; Col. 1:16; Heb. 1:2), and that the miracles He performed while on earth were intended to reveal His true nature and glory (John 1:14, 2:11, 20:31), it is deeply instructive to note that these works all involved *sudden transformations.* Thus, while it has been claimed by one philosopher that there is "no strategy as slippery and dangerous as analogy," the Biblical analogy of Christ's creative work in Genesis and in the Gospels remains irresistibly powerful.

In response to the mere word of Jesus Christ, for example, a raging storm *suddenly* ceased, a large supply of food *suddenly* came into existence, a man born blind *suddenly* had his sight restored, a dead man *suddenly* stood at the entrance of his tomb. Of the vast number of healing miracles performed by Christ, the only recorded exception to instantaneous cures is that of the blind man whose sight was restored in two stages, each stage, however, being instantaneous (Mark 8:25).[35] Such miracles were undeniable signs of supernaturalism in our Lord's public claim to Messiahship, and we may be quite sure that if, in His healing of the sick and crippled and blind, He had exhibited "the prodigal disregard for the passing of time that marks the hand of him who fashions a work of art,"[36] no one would have paid any attention to His claims! If the Sea of Galilee had required two days to calm down after Jesus said, "Peace, be still," the disciples would neither have "feared exceedingly," nor would they have "said one to another, 'What manner of man is this,

---

[35]This remarkable exception to our Lord's many thousands of instantaneous cures certainly cannot be used as a basis for progressive creationism, with its step-by-step (and yet supernatural) concept of origins. The very fact that this exceptional case is singled out in Scripture serves as a warning against those who would assume that this was God's basic method of creating the world. The Genesis account gives absolutely no hint of a multi-stage "creation" that continued throughout vast ages. The fact that the moon is said to have been brought into existence on the fourth day, with no reference to previously existing materials, calls for an *ex nihilo* creation, not a multi-stage creation. To borrow our Lord's comment from another context, "... if it were not so, I would have told you ...." (John 14:2).

[36]Leonard Verduin, "Man, A Created Being: What Of An Animal Ancestry?" *Christianity Today* 9:17 (May 21, 1965), 10. Verduin writes from a theistic evolution perspective.

that even the wind and the sea obey him?' " (Mark 4:39-41).

The profound theological implications of these facts for the Christian understanding of the origin of the moon and the rest of the universe can be recognized in the comment of a prominent British evolutionist:

> The theologian attributes certain *infinite* properties to his God; he is described as omnipotent, omniscient, and of infinite goodness. Now the Mind which reveals itself in the development of life on this planet is clearly not omnipotent, otherwise it would have assembled perfectly designed organisms directly from the dust of the earth without having to go through the long process of trial and error which we call evolution.[37]

Every effort to modify the suddenness and supernaturalness of creation events to make them more acceptable to the "modern mind" only results in the long run in minimizing and obscuring the true attributes of the God of creation. This has been a difficult lesson for many Christians to learn.

In the third place, the fact that God's work of creation was completed in six literal days clearly demonstrates that the creative work of each day was sudden and supernatural. In view of the widespread resistance to this concept, even in some Christian circles, it may be surprising to many people to learn how strong are the Biblical arguments in its support, if the indispensable historical/grammatical system of Biblical hermeneutics be accepted. Four of these arguments as well as answers to major objections will now be presented.

(1) Although the Hebrew word for "day" (*yôm*) can refer to a time period longer than 24 hours if the context demands it (e.g., "day of the Lord"), its attachment to a numerical adjective restricts its meaning to 24 hours ("first day," etc., with a precise parallel in Numbers 7:12-78).[38] The expression "one day" in Zechariah 14:7, claimed by some to be an

----

[37] John L. Randall, *Parapsychology and the Nature of Life* (London: Souvenir Press, 1975), p. 235.

[38] Technically, the description of a literal day as a period of 24 hours simply raises the question of the length of an "hour." In a personal communication, Harold L. Armstrong, editor of the *Creation Research Society Quarterly*, observes that "if 'hour' means one twenty-fourth of a day, any day is one of twenty-four hours. Why not specify 'day' as is done in Genesis: it is a period of darkness followed by a period of light."

exception, probably refers to a literal day also, because of the term "evening" in the same verse. Four times within the creation narrative the word "day" refers to the 12-hour period of daylight (1:5,14,16,18), but no numerical adjectives are used and the context clearly shows which sense is to be understood (which is true in English as well). Since the expression "in the day" (*beyôm*) in Genesis 2:4 lacks the numerical adjective, it could refer either to the first day or to the entire creation week. Better still, the phrase should simply be translated "when."[39] One opponent of the literal-day interpretation inadvertently displays the weakness of his position by admitting that Genesis 1 is the only place in the Bible where long "days" are numbered:

> If God created in six chronological days of indefinite duration, it is very likely he would number the days chronologically. This would then be the one place in the Bible which refers to chronological long period days, and, logically, the one place in which long period days are numbered.[40]

(2) The qualifying phrase, "the evening and the morning," which is attached to each of the creation days throughout Genesis 1, indicates a 24-hour cycle of the earth rotating on its axis in reference to any fixed astronomical light source (not necessarily the sun). The same phrase appears in Daniel 8:26 (cf. 8:14, ASV, NASB) where it must be understood literally. Some have claimed Psalm 90:6 as an example of a figurative use of the "evening" and the "morning." However, even this example is questionable, for the Genesis 1 formula is not used, and "morning" appears before "evening."

---

[39] Cf. Francis Brown, S. R. Driver and Charles A. Briggs, eds., *A Hebrew and English Lexicon of the Old Testament* (Oxford: Clarendon Press, 1975), p. 400. The term *beyom* with an infinitive construct lacks the chronological sharpness of *yom* with a numerical adjective. A total of 56 occurrences of the *beyom* usage is very strong attestation for the translation "when" or "at the time when" in Genesis 2:4.

[40] Russell W. Maatman, *The Bible, Natural Science and Evolution*, p. 95. Robert C. Newman and Herman J. Eckelmann admit, likewise, that the claim "that *yom* means a 24-hour day when used with *ordinal* numbers (1st, 2nd, 3rd, etc.), has the advantage that no clear counter-example can be cited with *yom* meaning a long period of time." They further admit that "the most common meanings of the words involved should be used in constructing a model." *Genesis One and the Origin of the Earth* (Downers Grove, Ill.: InterVarsity Press, 1977), pp. 61, 75.

Furthermore, even if "morning" and "evening" in Psalm 90:6 are used in a figurative sense, the figure would be meaningless if it did not presuppose the literal use of such terms in earlier historical narratives of Scripture, such as Genesis 1.

(3) A creation week of six indefinite periods of time would hardly serve as a valid and meaningful pattern for Israel's cycle of work and rest, as explained by God in the fourth commandment (Exod. 20:11, 31:17). While it is of course true that God *could* have created the world in six seconds or in six billion years if He had chosen to do so, such specula-tions are completely irrelevant in the face of the fourth com-mandment which informs us that God, as a matter of fact, chose to create the world in six days in order to provide a clear pattern for Israel's work periods and rest periods. The phrase "six days" (note the plural) can hardly be figurative in such a context. Leon Morris makes an interesting analogy between Genesis 1 and the seven successive days described in John 1:19–2:1.

> If we are correct in thus seeing the happenings of one momentous week set forth at the beginning of this Gospel we must go on to ask what significance is to be attached to this beginning. The parallel with the days of creation in Gen. 1 suggests itself, and is reinforced by the "In the beginning" which opens both chapters. Just as the opening words of this chapter recall Gen. 1, so is it with the framework. Jesus is to engage in a new creation. The framework unobtrusively suggests creative activity.[41]

(4) Since the word "days" in Genesis 1:14 is linked with the word "years," it is quite obvious that our well-known units of time are being referred to, their duration being deter-mined not by cultural or subjective circumstances, but by the fixed movements of the earth in reference to the sun. Other-wise the term "years" would be meaningless. We must assume that the first three days of creation week were the same length as the last three astronomically fixed days, because exactly the same descriptive phrases are used for each of the six days (e.g., numerical adjectives and the evening/morning formula) and all six days are grouped together in Exodus 20:11 to serve as a model for Israel's work-week. The fact

---

[41] Leon Morris, *The Gospel According to John* (Grand Rapids: Wm. B. Eerd-mans Publishing Co., 1971), p. 130.

that the sun was not created until the fourth day does not make the first three days indefinite periods of time, for on the first day God created a fixed and localized light source in the heaven in reference to which the rotating earth passed through the same day/night cycle.[42] To interpret the word "day" in this chapter as a long or indefinite period of time is thus completely arbitrary. The universe was created by God within one literal week.

In opposition to the literal-day interpretation, it has been asserted that Eskimos have a 6-month day instead of 24-hour day, so that even on the earth the length of the day is extremely flexible.[43] This argument is not valid. Even during winter months Eskimos can observe from the alternating phases of light on the horizon that a day lasts about 24 hours. Pressed to its logical conclusion, such an argument would mean that the word "day" could *never* mean a 24-hour period, either in the Bible or in our contemporary experience.

It has also been maintained that other passages of the Bible speak of a day in God's sight being as a thousand years. It is true that the Bible makes this statement (cf. Ps. 90:4; 2 Peter 3:8); but so far from destroying the literal-day interpretation of Genesis 1, it actually helps to establish it. In 2 Peter 3:8, for example, we are *not* told that God's days last a thousand years each, but that "one day is with the Lord *as* a thousand years." To say "as a thousand years" is a very different matter from saying "*is* a thousand years." This point has often been overlooked. If "one day" in this verse really means a long period of time, then we would end up with the following absurdity: "a long period of time is with the Lord as a thousand years." But a thousand years would be a long

---

[42] This light could not have been light from God's divine nature, for it was created ("let there be"). Furthermore, if it was the light of God Himself, half of the earth would not have remained in darkness. Thus, it must have been a created light, localized somewhere in the universe, and perhaps incorporated into the stellar universe on the fourth day of creation week, after its unique and temporary function was accomplished. It could not have been the sun, for reasons discussed in note 31 above. See the excellent discussion in H. C. Leupold, *Exposition of Genesis* (Columbus, Ohio: The Wartburg Press, 1942), pp. 51-53.

[43] Addison H. Leitch, *The Creation of Matter, Life, and Man* (fifth in a series of essays published by *Christianity Today*, September 16, 1966), p. 10.

period of time for human beings too!

The obvious teaching of 2 Peter 3:8 (as well as Ps. 90:4), then, is that God is above the limitations of time. One valid deduction from this fact is that *God can accomplish in one brief, literal day what man could not accomplish in a thousand years, if ever.* This is one of the astounding messages that comes through to us from the creation narrative of Genesis 1: *God has infinite power; we do not!* The prophet Jeremiah understood this: "Ah Lord God! behold, thou hast made the heaven and the earth by thy great power and stretched out arm, and there is nothing too hard for thee" (Jer. 32:17). Did the Creator, perhaps, really need the seventh day to rest from six days of creative work? The answer comes back with overwhelming clarity:

> Hast thou not known? hast thou not heard, that the everlasting God, the Lord, the Creator of the ends of the earth, fainteth not, neither is weary? .... He giveth power to the faint; and to them that have no might he increaseth strength (Isa. 40:28-29).

There is, in fact, simply no way for the human mind to grasp the power of God: "To whom then will ye liken me, or shall I be equal? saith the Holy One . . . . For as the heavens are higher than the earth, so are my ways higher than your ways" (Isa. 40:25; 55:9).

This enormously significant truth concerning God is shattered when one tries to stretch the account of creation to incorporate vast ages of time in order to make the passage more "reasonable," and thus to accommodate it to man's finite, uniformitarian level of thinking.[44] To twist the Scrip-

---

[44] A recent volume by Davis A. Young illustrates the tensions that are created when uniformitarian views of geology are confronted with the clear statements of Genesis 1. While insisting, on the one hand, that Scripture is far more authoritative than science (*Creation and the Flood: An Alternative to Flood Geology and Theistic Evolution* [Grand Rapids: Baker Book House, 1977], pp. 18-22), Young feels, on the other hand, that "if it can be demonstrated beyond all doubt that Scripture demands a 24-hour view of the days, then the Christian scientist must accept that and, in effect, give up geological science and turn to something else. If he is consistent in his faith in Scripture, he must do this" (p. 82). "Naturally," Young continues, "as a geologist, I am quite delighted with [the day-age view of Genesis 1] for I have become accustomed to thinking in terms of billions of years" (p. 91). The figurative view of Genesis 1 "gives the scientist great freedom" (p. 87), and leaves him "unfettered" (p. 113). "In short the Christian ought to be willing to let science advance in its own way and in its own time, that is, to

ture is to distort God's message to us. What the Apostle Peter said concerning Paul's letters is therefore completely applicable to the opening chapters of the Bible, ". . . in which are some things hard to be understood, which they that are unlearned and unstable wrest, as they do also the other scriptures, unto their own destruction" (2 Peter 3:16).

Another widely held objection to the literal-day interpretation of Genesis 1 is that the seventh day never terminated, for God is still resting from His work of creation (cf. Heb. 4:3-11).[45] But this argument introduces much confusion between historical events and their spiritual application. The "rest" of Hebrews 4 is primarily the spiritual rest of salvation (cf. Matt. 11:28-30) whereby the believer shares in the eternal blessedness and fulfillment which characterizes God.[46] Certainly God did not have to wait until the end of the sixth day of creation week for this kind of rest to begin! Thus, the first Sabbath was not instituted for God's benefit (cf. Isa. 40:28, John 5:17) but rather for man's benefit (Mark 2:27). *It is this often neglected point that is crucial in determining the duration of the original Sabbath day.*

How long, then, did the first Sabbath continue? It is obvious that all Israelites, to whom Sabbath observance was specifically applied by God, understood this period to be

---

develop naturally as new discoveries are made. One cannot force scientific thinking to advance along a particular line" (p. 114). How this approach honors the supremacy of Scripture is not convincingly explained. For further analysis of Davis Young's position, cf. J. C. Whitcomb, "The Science of Historical Geology," *The Westminster Theological Journal* 36:1 (Fall, 1973), 65-77.

[45]Cf. Russell W. Maatman, *The Bible, Natural Science, and Evolution*, pp. 90-94; cf. Newman and Eckelmann, *Genesis One and the Origin of the Earth*, pp. 65,85.

[46]Cf. Homer A. Kent, Jr., *The Epistle to the Hebrews* (Winona Lake, Ind.: BMH Books, 1972), p. 82, note No. 32: "[The fact that no terminus is mentioned for the seventh day] does not imply that the seventh day was not a literal day with an evening and a morning, just as the previous six days of creation. However, the author has used the silence of Scripture on this point to illustrate his argument that God's sabbath rest has never ended. The same method of argument is used in 7:3 regarding Melchizedek's absence of recorded birth, parentage, or death." Edward J. Young concludes that "the seventh day is to be interpreted as similar in nature to the preceding six days. There is no Scriptural warrant whatever (certainly not in Hebrews 4:3-5) for the idea that this seventh day is eternal." *Studies in Genesis One* (Presbyterian and Reformed Publishing Co., 1964), p. 77 note 73.

exactly 24 hours in length, based on the pattern of God's creation Sabbath:

> Six days shalt thou labour, and do all thy work: But the seventh day is the sabbath of the Lord thy God: in it thou shalt not do any work.... For in six days the Lord made heaven and earth, the sea, and all that in them is, and rested on the seventh day: wherefore the Lord blessed the sabbath day, and hallowed it" (Exod. 20:9-11).

Any Israelite who decided to extend his Sabbath observance indefinitely on the assumption that God's Sabbath still continues would have starved to death (cf. Exod. 35:3). Equally significant is the deduction that Adam and Eve must have lived through the entire seventh day of creation week before God drove them out of the garden, for God would not have cursed the ground (Gen. 3:17) during the very day He "blessed" and "sanctified" (Gen. 2:3).[47]

In conclusion, there is simply no escaping the fact that God intends us to understand the creation of the astronomi-

---

[47] For additional supporting arguments for a literal creation week, see Louis Berkhof, *Systematic Theology* (Grand Rapids: Wm. B. Eerdmans Pub. Co., 1955), pp. 150-64; Henry M. Morris, *Scientific Creationism*, pp. 221-30; Weston W. Fields, *Unformed and Unfilled* (Nutley, N.J.: Presbyterian and Reformed Publishing Co., 1976), pp. 165-79. It has recently been argued that the sixth day of creation must have lasted much longer than 24 hours because God must have given Adam enough time to become lonely. This is supposedly confirmed by the fact that when he was awakened and presented with Eve, he exclaimed: "This at last [*happa'am*] is bone of my bones" (Gen. 2:23a, RSV). Cf. Newman and Eckelmann, *Genesis One and the Origin of the Earth*, p. 131. But surely the term *happa'am* ("this once" or "now at length," cf. Brown, Driver, and Briggs, *Hebrew and English Lexicon of the Old Testament*, p. 822, under 2d) cannot be pressed to mean an absolutely (instead of a relatively) long period of time. Jacob could appropriately have used this expression after two or three hours of intense wrestling (Gen. 32:24). But it is not necessary to speculate about the possible uses of the term, for in Genesis 18:32 Abraham uses it at the end of a single conversation with his Lord! The authors explain that "here the strong emotional climax may build quickly because Abraham is bargaining with God" (*ibid.*, p. 133). But this destroys the entire argument, for Adam would hardly have been less emotionally involved with God and Eve. Those who maintain that Adam could not have named the birds and animals in one day with God's special help and with a freshly created, unfallen mind, must indulge in the typically hazardous uniformitarian extrapolations that characterize so many contemporary studies in cosmogony. Giving a little more time to Adam to name the animals may seem to be a very inconsequential issue. What Newman and Eckelmann are really asking for, however, is a "creation week" that lasted "15 billion to 20 billion years" with the seventh day yet future! (*ibid.*, pp. 83-85).

cal bodies, including the moon, to have been instantaneous. The implications of this profound fact with regard to currently popular attempts to harmonize Genesis with cosmic evolutionism should be perfectly obvious. To suggest "a gradual creation" of the moon may be conceivable to some minds. But for most people, such a concept would raise the very serious question as to whether God, as a matter of fact, ever created the moon at all. When the stupendous fact begins to dawn upon us, however, that the moon was created *instantaneously* and *ex nihilo*, all serious questions concerning the deity, omnipotence, and glory of the moon's Creator evaporate.[48] This is why the Hebrew/Christian approach to creation is shocking and transforming in its impact upon the human mind.

---

[48]The absolute date for the creation of the earth and moon cannot be determined Biblically because of the flexibility of the term "begat" in the genealogies of Genesis 5 and 11. There may have been additional patriarchs not included in these lists, just as three names were intentionally omitted in Matthew 1:8. However, the creation of the earth could hardly have occurred more than ten thousand years ago from the Biblical perspective. For a full discussion, see Whitcomb and Morris, *The Genesis Flood*, pp. 474-89. See also, J. C. Whitcomb, *Chart of the Period from the Creation to Abraham* (Winona Lake, Ind.: BMH Books [Box 544], 1976). The relative date for the moon's creation is Biblically fixed three days after the earth's creation and two days before man's creation (see discussions elsewhere, pp. 72 [note 31], 153-55).

CHAPTER IV

Lunar Geology

"The empirical basis of objective science has nothing 'absolute' about it. Science does not rest upon rock bottom. The bold structure of its theories rises, as it were, above a swamp. It is like a building erected on piles. The piles are driven down from above into the swamp, but not down to any natural or 'given' base; and when we cease our attempts to drive our piles into a deeper layer, it is not because we have reached firm ground. We simply stop when we are satisfied that they are firm enough to carry the structure, at least for the time being."

Karl R. Popper
(The Logic of Scientific Discovery,
1959, p. 11)

The moon, like most other solar system bodies, is an object of great geological extremes. *Lunar mountain ranges* first identified by Galileo extend above the surrounding plains with lofty peaks, some surpassing Mount Everest in height. Especially high elevations near the northern lunar pole, the so-called "mountains of eternal light," withdraw from clear sunshine only at times of a lunar eclipse.

Water, so important in shaping the earth's surface, is virtually nonexistent on the moon. The inevitable processes of material breakdown and increasing disorder on the lunar surface result mainly from thermal expansion affects. The lack of a substantial lunar atmosphere results in a 132°C (270°F) temperature plunge each time the sun sets. As a result lunar rocks crack and break down from stresses set up by temperature-induced expansion and contraction. On a much smaller scale, there is similar "mechanical" weathering of rocks on earth due to the solar heat. The slow movement of lunar material toward valley bottoms is accomplished by ground tremors (moonquakes) and impacts of small meteorites.

While terrestrial impact craters are frequently destroyed by erosion, *lunar craters* are often well preserved. Clavius, a large crater on the moon's visible side, measures 340 kilometers (211 miles) across. Oceanus Procellarum, thought by some to be a crater remnant, exceeds the Mediterranean Sea in area. Smaller craters within craters cover the moon like an intense battlefield. There are estimated to be 200,000 craters with diameters larger than a kilometer on the entire moon.[1] Some of these craters resemble volcanic formations with peaks rising in the center. Others are so large and deep that meteorite impact is thought to be required.

Besides mountains and craters, many other distinctive surface features of the moon can be seen with a telescope. *Maria* (pronounced mah'-re-ah; singular: *Mare*), the Latin word for seas, cover one-half of the moon's visible side, resembling large, dark, somewhat circular patches. Bright *rays* appear to radiate outward hundreds of miles from some of the larger craters. They are considered to represent debris tracks deposited from the ejecta of major meteorite impacts. With a lunar gravitational field which is only 17% as strong as that of earth, projectiles have much longer trajectories. The Apollo explorers as well could have bounded for several meter distances on the moon, but were hindered by bulky space suits. Sinuous lunar *rilles* which meander across the moon's surface are probably collapsed lava draining channels, since the lack of water rules out river channels. The visible cracks and valleys are apparently adjustments in the moon's surface to stress from heating and tidal pull.

LUNAR ROCKS

The Apollo teams returned to earth with an abundance of

---

[1]S. P. Wyatt, *Principles of Astronomy* (Boston: Allyn and Bacon, Inc., 1977), p. 152.

lunar memories, photographs, and material samples. Suddenly the exploration of the moon's crust became possible with electron microscope analysis of samples, as well as with the faithful telescope. Bringing lunar materials into the laboratory provided an unprecedented experimental increase in resolution by ten trillion times over earth-based observation of the moon.[2]

The returned rocks resemble earth varieties in some respects and differ significantly in other ways. The three varieties of collected samples are *crystalline rock, soil,* and *breccia* (Table IV-1). The geological term *regolith* is given to the general lunar rubble pile of dust, pebbles, and boulders.

Crystalline Rock

The *crystalline rocks* contain the same chemical elements as earth rocks and established terms can be used to describe them. Basalt, similar to the terrestrial volcanic rock, cooled from molten lava, and is especially common in the maria. It is a fine-grained igneous rock containing principally the mineral groups plagioclase (solid solution of $Na[AlSi_3O_8]$ and $Ca[Al_2Si_2O_8]$) and pyroxene (a large group of iron-magnesium silicates). The identifying small crystals suggest a rapid cooling of the moon's surface in the past. All moon

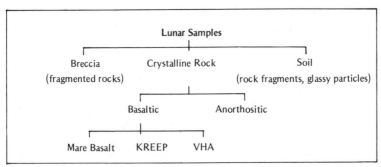

Table IV-1. A Classification Diagram of returned Lunar Samples.

[2]With the Palomar telescope, lunar detail can be resolved (seen clearly) down to a fraction of a kilometer (1000 meters). With the field ion microscope, individual atoms one Angstrom in size (10^{-10} meters) can be photographed. See K. F. Weaver, "Electronic Voyage through an Invisible World," *National Geographic,* 151:2 (February, 1977), 274-90. The increase in resolution resulting from bringing lunar samples into the laboratory is $10^3/10^{-10} = 10^{13}$, or ten trillion times.

rocks contain a proportionately higher amount of heat-resistant elements such as calcium, aluminum, titanium, zirconium, and chromium than rocks from the earth. Conversely the volatile (easily vaporized) elements sodium, potassium, and lead are relatively depleted on the moon. The lunar basalts subdivide into three varieties: mare basalt, "KREEP" basalt, and "VHA" basalt. Mare basalt is the common floor covering of the maria. "KREEP" basalt is a less common potassium-rich form found in Mare Imbrium and Oceanus Procellarum. The strange name comes from the sample's high content of potassium (K), rare earth elements (REE— europium, terbium, and lutetium), and phosphorus (P). The "VHA" basalts are the nickname for Apollo 16 samples from the Descartes highlands which are very high in aluminum. Because basalt is dark colored, it imparts a dark hue to the lunar maria.

A second important lunar rock is anorthosite. It is composed almost completely of calcium-rich plagioclase and forms the dominant rock of the lunar highlands. Because of its lack of pyroxene, anorthosite is light gray, making the highlands lighter in color than the basalt-covered maria. The color contrast between anorthosite and basalt is responsible for the appearance of illusion of a human face ("man-in-the-moon") on the moon's near surface.

Lunar Soil

The sterile lunar *soil* consists of the powdered remains of collisions between meteorites and the igneous surface. Unlike the typical soil of earth, it contains no organic matter and virtually no moisture. Small bright beads of colored glass give variety to the soil. Such beads may indicate the melting of material during impacts and subsequent rapid cooling. To this lunar soil are added daily streams of hydrogen, helium, and a variety of ions and atoms from the solar wind, a tenuous ionized gas ejected from the sun.

Breccia

Brecciated or broken samples are fragmented rocks in which constituent particles are sharply angular rather than rounded as in the case of terrestrial conglomeratic rocks. The particles are composed of small rock fragments, glass, and soil compacted into cohesive aggregates. These breccias, like the glass beads, may be due to shock melting during the impact of meteorites on the lunar surface. However, evidence for this impact hypothesis is limited.[3] Since the pebbles making up the breccias appear to come from remote sites, they cannot be fused at the new site by the same shock that fragmented them in the first place. Violent collision is not ruled out as a *possible* cause, but it is not a certainty.

Lunar Rock Analysis

Several new minerals were found on the moon (Table IV-2). This is really not surprising since even the list of identified terrestrial minerals presently numbers in the thousands, and is still growing. Of greater interest are the familiar earth components which the moon lacks: granite and sedimentary rock, moisture, carbon, and free oxygen.

The earth's igneous crust is predominantly granite (the continental crust) and basalt (the oceanic crust). Sedimentary layers formed by water deposition cover 74% of the earth's land surface. Metamorphic or changed rocks form a minor constituent of the earth's crust. However, on the moon igneous rock is found almost exclusively. No coarsely crystalline granitic rocks have been found, nor any sedimentary deposits which would hint at a former water supply.[4]

Lunar Mineral	Chemical Symbol
Armalcolite	$(Fe,Mg) Ti_2O_5$
Tranquillityite	$(Fe, Y, Ca, Mn) (Ti, Si, Zr, Al, Cr) O_3$
Pyroxferroite	$Ca Fe_6 (Si O_3)_7$

Table IV-2. Three New Igneous Rock Minerals Found on Mare Tranquillitatis by Apollo 11. Note that the first mineral name is an acronym for the Apollo team of <u>Arm</u>strong, <u>Al</u>drin, and <u>Col</u>lins.

---

[3] S. Tolansky, "Lunar Glass: Interferometric Evidence for Low-Temperature Shock," *Science* 176:4035 (May 12, 1972), 671. The lack of mechanical shock evidence of lunar breccias is discussed.

[4] Evidence of local sedimentation has been found in scattered places on the moon. However, they are probably reversed deposits from meteoric impact.

Neither is there more than a trace of water within the lunar rocks. Almost all earth rocks, on the other hand, contain 1-2% water. This lack of lunar moisture rules out many varieties of clay minerals on the moon, just as the absence of oxygen rules out iron oxides. The result is that the earth contains 20 different minerals for every one found on the moon thus far.

Because of early fears of possible infecting microbes in the unknown lunar environment, the first astronauts and their samples were temporarily placed in isolation chambers upon returning to earth. However, the lunar rocks were determined to be barren of life, so the quarantine period was dropped completely following the third Apollo landing. The lack of lunar water and life was a disappointment to those who had hoped the moon would support the Darwinian concept of the natural evolution of life, as *Pictorial Astronomy* indicates:

> None of the rock or soil samples that have been returned show any evidence of microscopic living, previously living, or fossil material. This was a disappointment to those advocating widespread life in the universe.[5]

The lack of life-building carbon and free oxygen also quenched evolutionary optimism concerning the moon. Returned lunar rocks contain only 'parts per million' traces of carbon and carbon compounds which themselves may be due to contamination by man. The oxygen deficiency is responsible for the small amounts of pure metallic iron that have been found on the moon. On earth this iron would be quickly oxidized to rust. Hence the returned iron samples must be permanently stored in a dry nitrogen atmosphere to maintain their lunar form. One cannot help but reflect on the extreme differences between the comfortable earth and its important but entirely inhospitable satellite companion. If the earth were without its protective atmosphere and abundant water supply, the terrestrial environment would be similar to that of the moon![6] The contrast of the moon, or any other of the

---

[5] D. Alter, C. H. Cleminshaw and J. G. Phillips, *Pictorial Astronomy*, (New York: Thomas Y. Crowell Company, 1974), p. 265.

[6] Scripture has much to say concerning the earth's water supply and its life-sustaining cycle. See Deut. 11:11, Job 12:15, Isa. 55:10, Psa. 135:7, and Eccl. 1:7.

planets, provides a vivid demonstration of God's care in providing a beautiful earth.

HYPOTHETICAL MOON HISTORY

From the fascinating variety of lunar features, an evolutionary history of the moon has been suggested. As outlined in Table IV-3, the proposed sequence of events is as follows:

1. The sun and all other solar system objects, including the moon, have an unknown origin approximately 4.6 billion years ago.
2. The outer layers of the moon are melted by radioactivity to a depth of 160 kilometers, during the first billion years.
3. Material differentiation occurs. Lighter matter rises, while heavier material settles.
4. Celestial debris causes intense cratering during the first billion years.
5. Lunar lava heated by internal radioactivity rises and fills in the large surface maria basins.
6. Lava flows ended approximately 3 billion years ago. A sprinkle of meteorites continues. The moon becomes quiet and "dead" through the present time.[7]

Although the listed events are often taken as a factual summary of the moon's history, they in truth only describe an uncertain, hypothetical model. Paul Gast, then Chief of the Planetary and Earth Sciences Division of the NASA Manned Spacecraft Center, gives an idea of the difficulty involved in deciphering the lunar past: "It's like having a giant jigsaw puzzle with the pieces all mixed up but no picture on the lid of the box."[8] In view of the past unknown changes which have taken place on the moon, a clear explanation of its origin from a naturalistic viewpoint becomes all the more uncertain, as Goldreich states:

> Speculation on the origin of the moon is largely concerned with deciding where along the calculated evolutionary path the moon actually originated. If there were some readily discernible

---

[7] J. A. Wood, "The Moon," *Scientific American*, 233:3 (September, 1975), 92-102. R. A. Pacer and W. D. Ehmann, "The Apollo Missions and the Chemistry of the Moon," *Journal of Chemical Education*, 52:6 (June, 1975), 350-356.

[8] "The Moon" *1971 Reader's Digest Almanac and Yearbook*, (Pleasantville, New York: Reader's Digest Association, Inc., 1971), p. 99. The quote is from an address presented by Dr. Gast in late 1970.

Date (Billions of Years Before the Present)

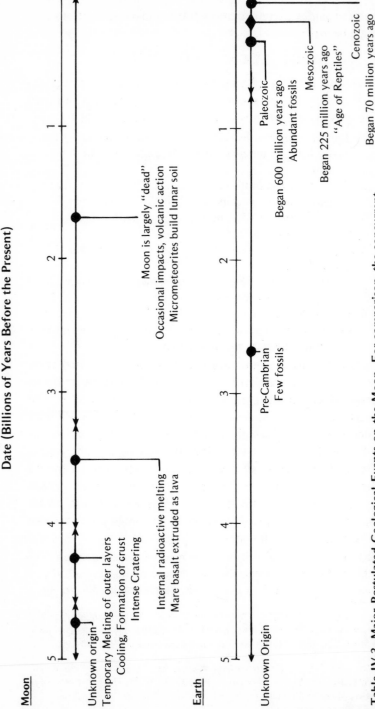

Moon

Unknown origin
Temporary Melting of outer layers
Cooling, Formation of crust
Intense Cratering

Internal radioactive melting
Mare basalt extruded as lava

Moon is largely "dead"
Occasional impacts, volcanic action
Micrometeorites build lunar soil

Earth

Unknown Origin

Pre-Cambrian
Few fossils

Paleozoic
Began 600 million years ago
Abundant fossils

Mesozoic
Began 225 million years ago
"Age of Reptiles"

Cenozoic
Began 70 million years ago
Present era

Table IV-3. Major Postulated Geological Events on the Moon. For comparison, the concurrent uniformitarian geologic periods for earth are also shown.[7]

landmark to which the evolutionary path pointed, there would undoubtedly be more agreement on the origin of the moon. In the absence of any landmarks a bewildering variety of hypotheses have been put forward.[9]

Table IV-3 shows that the assumed history of the moon covers billions of years. However, evidence for this is limited to uncertain radiometric dating interpretations. This long-age hypothesis is similar to the earth's assumed lifetime, which is also based upon radiometric results, together with a vast-age view of the evolution of plants and animal life, sedimentation and erosion rates, etc. The long ages suggested for both the earth and moon are really not surprising since radiometric methods commonly used for rock are not sensitive to ages less than many thousands of years. This is due to the very slow decay of the radioactive isotopes used (see p. 98). In truth, the uncertainties of initial conditions and environment may render radiometric dating meaningless over eons of time. The method often appears to be mainly "window-dressing," used to give a measure of scientific credibility to the uni-formitarian geological time scale. On a much shorter time-scale of hundreds or a few thousand years, radiometric dating using the isotope carbon-14 has been valuable, when independent dating methods are available as a cross-check. The uncertainties of the radiometric method arise with the very long-lived isotopes, for which dating results involve an extrapolation far back in time with no way to verify the results with an independent method. Note that even the "long-age" radiometric investigations are by no means *worthless*, but are entirely *uncertain* at this time. When properly interpreted, radioactive isotope abundances will tell much about atomic migration, exchange, and contamination in the lunar and earth environment. These are complex and formidable problems in present solid-state science.

It is evident that lunar exploration began and ended with a firm commitment to long ages, so strongly emphasized in earth science, as Owen Gingerich implied: "Astronomy students unfamiliar with the geologists' time scale ought to

[9]P. Goldreich, "Tides and the Earth-Moon System," *Scientific American*, 226:4 (April, 1972), 50.

memorize the order of the five geologic periods."[10] This adherence to a uniformitarian view is recognized by many as having little basis. John W. Wells of Cornell University writes:

> Absolute age determinations of points on the geological time scale based on radioactive decay are generally accepted as the best approximations, even though these rest on a series of assumptions, any one of which may be upset at any time. At present there is no means of confirming or denying the accuracy of these determinations by independent methods. ... The most recent estimates of geological time based on rates of radioactive decay give the Cenozoic era a length of 65 million years, the Mesozoic 165 million years, the Paleozoic 370 million years, and so forth. The beginning of the Cambrian period is placed about 600 million years ago. To accept these figures is an act of faith that few have the temerity to refuse to make.[11]

Thus, the common view is that three billion years ago, after a violent period, the moon simply died and stopped changing except for the gentle rain of micrometeorites and solar wind particles, becoming in effect the "museum of the early solar system."[12] A NASA publication expresses this idea:

> If men had landed on the Moon a billion years ago, it would have looked very much as it does now. The surface of the moon now changes so slowly that the footprints left by the Apollo astronauts will remain clear and sharp for millions of years.[13]

The view of an immense time scale gives rise to many paradoxes, three of which will be discussed.

First, the moon has not accumulated the amount of dust that some scientists believe eons of time would provide. Be-

---

[10]*Frontiers in Astronomy* (Readings from *Scientific American*), ed. by O. Gingerich, (San Francisco: W. H. Freeman and Co., 1970), p. 3.

[11] J. W. Wells, "Paleontological Evidence of the Rate of the Earth's Rotation," *The Earth-Moon System*, ed. by B. G. Marsden and A. G. W. Cameron (New York: Plenum Press, 1966), 72.

[12]Scientists at the University of Chicago have recently announced that the universe is 20 billion years old, based on the radioactive isotope rhenium 187, whose half-life is 40 billion years. See "20-billion-year universe," in *Science News*, 111:14 (April 2, 1977), 215.

[13]B. M. French, *What's New on the Moon?* (Washington, D.C.: NASA, 1976), p. 15.

sides the constant breakdown of surface material due to thermal weathering, dust from comet meteoric debris and material from the sun itself are constantly filtering onto the surfaces of both the earth and moon. Earth dust is washed into the seas, and moon dust accumulates in any low-lying area. In the early history of the solar system, dust concentrations should have been even greater than at present, based on uniformitarian assumptions. It is little wonder that pre-Apollo concerns included the possibility of a landing craft being swallowed up by an unstable dust-filled surface, particularly on the smooth mare landing sites.[14] Of course, lunar probes eventually showed a very thin coat of dust on the moon, barely sufficient to show tracks. The hazard of sinking into a sea of dust on the moon was forgotten. Efforts have been made to re-evaluate dust accumulation rates and to find a mechanism for lunar dust compaction,[15] but the dust's absence remains unexplained in view of the uniformitarian billion year time scale.

A *second* problem with a five-billion-year interpretation of the moon's history is its present instability. It is by no means a "dead" or "fossilized" satellite. Much evidence pointing to

---

[14]The British astronomer Raymond A. Lyttleton predicted a layer of moondust several miles in thickness. See R. A. Lyttleton, *The Modern Universe* (New York: Harper and Brothers, 1956), p. 72. "Gold (1955) proposed that vast amounts of mobile dust migrated and collected in low places on the moon to make the flat lunar plains, and Shoemaker (1965) anticipated that it would be 'tens of meters deep.' " William K. Hartmann, *Moons and Planets* (Belmont, California: Bogden and Quigley, Inc., Pub., 1972), p. 280. These references by Hartmann are to Thomas Gold, "The Lunar Surface," *Monthly Notices of the Royal Astronomical Society* 115 (1955), 585, and to E. M. Shoemaker, "Preliminary Analysis of the Fine Structure of the Lunar Surface," *Ranger VII, Part II Experimenter's Analyses and Interpretations JPL TR 32-700*, 75. Pre-Apollo calculations by Isaac Asimov predicted a depth of lunar dust of *at least* 50 feet. Cf. I. Asimov, "14 Million Tons of Dust Per Year," *Science Digest*, 45:1 (January, 1959), 33-36. Jay Pasachoff writes, "there were those who thought it [the moon] might be powdery, that a spacecraft would sink and would never be heard of again . . . It was only the soft landing of the Soviet Luna and American Surveyor Spacecraft on the lunar surface in 1966 and the photographs they sent back that settled the argument over the strength of the lunar surface; the Surveyor perched on the surface without sinking in more than a few centimeters. J. H. Pasachoff, *Contemporary Astronomy* (Philadelphia: W. B. Saunders Co., 1977), pp. 294-5.

[15]T. Mutch, *Geology of the Moon* (New Jersey: Princeton University Press, 1972), pp. 256-57.

a moon which is still actively adjusting to its original state of
creation is treated in detail in the following chapter on transient phenomena. Such findings as these cast a shadow of
doubt on the geological time scale as presented in Table IV-3.
Should the dates be scaled down from billions to thousands
of years? As one might expect, evidence also exists for a
vastly younger earth,[16] and solar system,[17] than is generally
believed to be possible.

A *third* area of unexpected results involves the dating of
lunar soil, assumed to be the product of the surrounding rock
environment. Rubidium-strontium radiometric dating shows
that in many cases the soil is "a billion years older" than the
adjacent rocks, as a data summary states:

> An Apollo 11 soil sample was 4.6 billion years old. This moon
> dust was about a billion years older than the chunks of rock that
> lay strewn about it. Offhand this seems impossible, and it is be-
> wildering to scientists. . . . The natural assumption is that the soil
> is simply the powdered remains of the rock lying amidst it. But
> like so many other assumptions about the moon, this one simply
> does not hold up, since the rocks are a billion years younger.[18]

Lunar soil analysis leads to a second conflict with the uniformitarian idea that ancient physical processes on the moon
were no different from those observed today. The lunar surface does not reveal the extent of soil mixing that long ages
predict.

> Another study revealed that the top [lunar surface] layer of
> bismuth and cadmium remained at the surface for 15 million
> years. If the lunar soil is constantly being dug up by many small
> impacts, more mixing should have occurred.[19]

It is also true that rocks dating back to the assumed beginning of the moon have not been found. The radiometric

---

[16] J. C. Whitcomb, Jr. and H. M. Morris, *The Genesis Flood* (Philadelphia: The
Presbyterian and Reformed Publishing Company, 1961).

[17] H. Slusher, "Some Astronomical Evidences for a Youthful Solar System,"
CRSQ, 8:1 (June, 1971), 55-57.

[18] "The Moon," *1971 Reader's Digest Almanac and Yearbook*, 99.

[19] "At the Moon Conference: Consensus and Conflict," *Science News*, 99:4
(January 23, 1971), 62.

The plaque on the Eagle's landing stage is engraved: "Here men from the planet earth first set foot on the moon, July 1969, A.D. We came in peace for all mankind." (NASA)

Sampling the lunar surface material. (NASA)

A photomicrograph of an Apollo 17 lunar sample thin section. Thin sections of earth rocks under polarized light show similar colors, which identify minerals within the sample. (NASA)

Another lunar thin section, brought from the Fra Mauro landing site of Apollo 12. (NASA)

Apollo 17 view of earth, taken during the final lunar landing mission in NASA's Apollo program. Almost the entire coastline of the African continent is visible. (NASA)

The 40-mile diameter crater Goclenius, showing rilles which cross the crater floor and rim. (NASA)

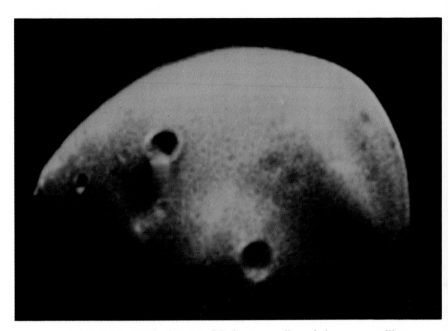

A computer-generated picture of Deimos, smaller of the two satellites of Mars. Resolution shows objects as small as 200 meters. Note the craters and nonspherical satellite shape. (NASA)

The planet Jupiter, from a distance of 2,600,000 km (1,615,000 miles), shows cloud markings, the Great Red Spot, and the shadow of Jupiter's moon, Io.

method readily gives ancient dates, but they still fall short of expectations, as W. K. Hartmann reports:

> Contrary to the hopes of scientists, astronauts found very few rocks actually dating back to the beginning. A few date from 4.3-4.0 eons [billions of years]. To go back to the very beginning we must rely on theory.[20]

Because of this lack of early rocks, there is no evidence that the lunar craters and mare lava regions were formed over an extreme length of time. The heat involved in the cratering and mare formation processes is said to have "reset" the uniformitarian radiometric clocks. That is, the apparent melting of the lunar surface destroyed any previous records.

A lengthy cratering period is countered by the fact that the earth-side of the moon is less cratered than the far side. Non-uniform cratering of Mars, Venus and Mercury as well as the moon might suggest a catastrophe such as a giant swarm of interstellar asteroids moving through the solar system. The earth has its share of craters as well, and appears to have undergone a similar early bombardment.[21] When could this have occurred? From the Biblical perspective, catastrophes of various kinds may have taken place in the universe any time following the primeval rebellion of mankind (cf. Gen. 3:14-24; Rom. 5:12; 8:20-22). It has been suggested that impacting objects were used in forming the earth's crust (not as a means of judgment) during the first three days of creation. More likely, widespread cratering of the earth and moon occurred at the time of the universal deluge that destroyed humanity and involved the collapse of the antediluvian vapor canopy and the upheaval of the "fountains of the great deep" (Gen. 7:11). Some of the smaller terrestrial craters and some of the lunar ray craters may have formed subsequent to the deluge. It must be stated, however, that no objective evidence for the absolute age of cratering has yet been dis-

---

[20] W. K. Hartmann, "The Moon's Early History," *Astronomy*, 4:9 (September, 1976), 9.

[21] For a discussion of large earth craters, see W. K. Hartmann, *Moons and Planets: An Introduction to Planetary Science* (Belmont, California: Wadsworth Publishing Company, Inc., 1972), p. 185; T. A. Mutch, *Geology of the Moon* (Princeton, New Jersey: Princeton University Press, 1972), p. 93; A. O. Kelly and F. Dachille, *Target: Earth* (Carlsbad, California: Target: Earth, 1953).

covered either in Scripture or in astronomical or geological science.

RADIOMETRIC DATING EVALUATION

Radiometric dating of lunar rocks involves several isotopes which can sometimes be cross-checked for agreement. Each method makes use of the spontaneous transformation of a parent nucleus into a stable daughter nucleus at a particular rate, designated by its half-life. Several well-known transitions are:

			Half-life in billions of years
U^{238}	Pb^{206}	(Uranium-Lead)	4.51
U^{235}	Pb^{207}	(Uranium-Lead)	0.713
Th^{232}	Pb^{208}	(Thorium-Lead)	14.1
Rb^{87}	Sr^{87}	(Rubidium-Strontium)	47
K^{40}	Ar^{40}	(Potassium-Argon)	1.31

By measuring the amount of daughter nuclei within a sample and subtracting an assumed original quantity of daughter nuclei based on isotope ratios in samples without the radioactive isotope, the time since the rock has solidified can be calculated. The technique represents an advanced state of chemical analysis. It requires extremely precise chemical and/or radiometric analysis for parent and daughter abundance ratios, and may involve the use of stable isotope ratios for normalization.

The end result of this major effort, unfortunately, may be ambiguous, and its interpretation is subject to increasing criticism in the scientific literature.[22] Lunar rocks are found to have a tremendously wide spread in ages. Some results indicate a sample age of 20 billion years. Other samples have an apparent age of only a few million years, mere "seconds" on the geologic time scale. Table IV-4 gives a summary of the published variation between different radiometric dating methods for twelve Apollo samples.[23] Of the hundreds of

---

[22]H. C. Dudley, "Radioactivity Re-examined," *Chemical and Engineering News* (April 7, 1975), 2; P. A. Catacosinos, "Do Decay Rates Vary," *Geotimes* 20:4 (1975), 11; R. V. Gentry, "Radiohalos in a radiochronological and cosmological perspective," *Science* 184:4132 (1974), 62-66.

[23]John G. Read (Scientific-Technical Presentations, P. O. Box 2384, Culver

other lunar samples that have been analyzed for age, many show better agreement. Still others probably show greater discrepancy. The variation in results shown by Table IV-4 is significant, and is understood when the inherent uncertainties and unknown initial conditions of rock formation are considered. The following is a partial list of events which may alter a sample radiometric calculation in the direction of longer apparent age than actual:

1. Inadequate allowance for the daughter nuclei initially present in the sample at the time of its formation.
2. Migration of daughter nuclei *into* the sample (or parent nuclei *out* of the sample) sometime after formation.
3. A possible shortening of the decay half-life during the sample's history.[24]

The rubidium-strontium method of lunar dating is particularly misleading. Rubidium, the parent nucleus, is a volatile element and can easily become heated sufficiently to move freely over the moon as a gas. Robert C. Wright, Senior Development Engineer with Princeton Applied Research Corporation, has calculated that the heat of just one lunar day is amply sufficient for the transfer of rubidium to or from a sample.[25]

City, California 90230) compiled the list of references for Table IV-4. The data is taken from the following sources:

Proceedings of the Second Lunar Science Conference, Vol. 2, pp. 1117, 1494, 1496, 1499, 1516, 1539, 1593, 1620, 1631 (1971).

Apollo 12 Preliminary Science Report (NASA SP-235), pp. 205-208 (1970).

Science, 167, 3918, pp. 462-463, 471-473, 479-480, 555-558 (1970).

Earth and Planetary Science Letters, 14, 281 (1972).

Proceedings of the Third Lunar Science Conference, Vol. 2, 1550 (1972).

Proceedings of the Fourth Lunar Science Conference, Vol. 2, pp. 1200 and 1804 (1973).

Earth and Planetary Science Letters, pp. 17, 36 (1972).

[24] D. B. DeYoung, "The Precision of Nuclear Decay Rates," in *CRSQ* 12:1 (March, 1976), 38-41. Here it is shown that at least twenty isotopes have shown decay half-life variation, including carbon-14. See also K. P. Dostal, M. Nagel, and D. Pabst, "Variations in Nuclear Decay Rates," *Zeitschrift für Naturforschung* 32a (April, 1977), 345-361.

[25] R. C. Wright, "Effects of Volatility on Rubidium-Strontium Dating," *Pensée* 2:2 (May, 1972), 20. For further discussion of the vaporization and redistribution of the volatile elements mercury, lead, potassium, and ruthenium on

| Apollo Sample No. | Ages in Billions of Years | | | Age inconsistencies, extremes in billions of years |
| | Uranium-Thorium-Lead Method | | Potassium-Argon Method | |
	Low	High		
10017	3.60	4.79	2.2	2.59
10057	3.96	4.17	2.3	1.87
10060	3.36	5.76	—	2.40
10084	4.31	8.20	>7	3.89
12070	3.63	4.50	>7	>3.37
12032	3.38	4.40	>7	>3.62
12063	3.75	4.09	2.6	1.49
12013	.7*	4.6	>6	>5.3
14310	5.3	11.2	—	5.9
14053	5.4	28.1	—	22.7
15426	4.6	16.2	—	11.6
66095	5.6	14.1	—	8.5

*Age determination using a Uranium-Thorium/Helium Technique

Table IV-4. Variation in Ages for Apollo Sample Material. Columns 2 and 3 list minimum and maximum dating results using any of the isotope ratios Pb^{206}/U^{238}, Pb^{207}/U^{235}, Pb^{207}/Pb^{206} or Pb^{208}/Th^{232}. Column 4 contains K^{40}/Ar^{40} data. The last column shows the largest possible discrepancy in sample age determination, found by subtracting the lowest value from the highest. The table does not list value uncertainties, which are typically given as 1-10%. The uncertainties would lead to overlap between methods in several cases. The symbol ">" means "greater

A concluding question arises as to how the radiometric clocks have been "reset," since lunar rocks are not as old as the assumed age of the moon. Either the past influx of matter had a correspondingly greater amount of the parent radionuclei which "just happened" to bring the parent-daughter ratio back to its initial value, or the melting caused nuclear fusion, reversing the decay which had occurred. This second choice involves a process so energetic that it is believed to occur only in certain extremely massive supernova explosions. In short, it is not clear by what method the previous radiometric record has been destroyed. One possible process may involve the third item in the previous list of radiometric dating alterations. But the reverse of number three (migration of parent nuclei *into* the material) is only possible if the parent was as much younger as the elapsed lunar time. That is, if the lunar rock ages were reset after one billion years, then the matter which did the resetting had to be fresh from a fusion process of the supernovae in order not to have accumulated an equal amount of daughter nuclei. If local enrichment is postulated, then the accuracy of the techniques used are certainly not valid to anywhere within a billion years.

---

the lunar surface, see "Third Lunar Science Conference," in *Science*, 176:4038 (June 2, 1972), 975-981.

The journal *Pensée* was dedicated to an expression of unpopular scientific views, such as those of Immanuel Velikovsky. In 1975, *Pensée* was replaced by the journal *Kronos*. Velikovsky has long supported unpopular ideas about the moon, including the belief that the moon was molten only 3000 years ago, and that lunar craters are due to the collapse of large bubbles. It should be realized that Velikovsky by no means believes in a recent origin or creation of the universe, only recent catastrophes: "Before we proceed, I wish to make it clear that the question is not *when* the rocks have been *formed* or for the *first* time crystallized, but when they were heated and partly molten for the last time. The *age* of the (moon) rocks is not in dispute, only the time of the 'carving' of the lunar surface. The rocks could be billions of years old" [italics are his]. Immanuel Velikovsky, "When Was the Lunar Surface Last Molten?" *Pensée* 2:2 (May, 1972), 20.

Many of Velikovsky's ideas on catastrophic phenomena are as dynamically improbable as the lunar capture hypothesis. His attitude toward Scripture is that the first chapter of Genesis is "a myth, a tale brought down from exotic and later sources." Immanuel Velikovsky, "Earth Without a Moon," in *Pensée* 3:1 (Winter, 1973), 25. Nevertheless, Velikovsky often appeals to Scripture as a record of ancient events.

The different methods of radiometric dating, when checked against each other, often are in approximate agreement. If the results are misinterpreted as to age, as proposed here, then a common unknown factor (a measurement or an assumption which is defective) may be perturbing all the age values to a longer apparent age than actual. Another explanation in some isolated cases of dating conclusions may be a "tracking phenomenon." By this is meant a tendency of reported scientific measurements to cluster about an incorrect value. Researchers are often reluctant to report findings too far different from previous results in their published findings. This clustering effect shows up in reports of nuclear half-life determinations, and it may also rule the 4.5 billion year assumed history of the earth and moon.[26] Studies of historical experiments have raised similar questions about some of the data recorded by such men as Claudius Ptolemy and Isaac Newton.[27] An alternative explanation for correlated radiometric dates may be the recent creation of the moon with built-in internal complexity (a rich variety of nuclei ratios).

The recent exploration of the moon has given rise to a detailed evolutionary view of lunar development, just as the popularization of the theory of continental drift has revolutionized the geological interpretation of earth history. Vast-age chronologies are assumed for both the earth and moon by most scientists. However, problems continually arise from this uniformitarian concept of eons of time. With respect to the moon, a sampling of radiometric dating discrepancies of rocks and soil has been discussed in this chapter. A detailed examination of other lunar geological controversies would fill several additional chapters. Is the interior of the

---

[26]Y. LeGallic, "Validity of Radioactive Standards," *Radioactivity Calibration Standards*, edited by W. B. Mann and S. B. Garfinkle, National Bureau of Standards Special Publication 331 (Washington, D.C.: U. S. Government Printing Office, 1970). See also a reference to the "reinforcement syndrome" by N. D. Watkins, "Review of the Development of the Geomagnetic Polarity Time Scale and Discussion of Prospects for its Finer Definition," in *Geological Society of America Bulletin* 83:551 (1972), 556.

[27]R. S. Westfall, "Newton and the Fudge Factor," *Science* 179:4075 (February 23, 1973), 751-758. R. R. Newton, *The Crime of Claudius Ptolemy* (Balti-(Baltimore: Johns Hopkins University Press, 1977).

moon hot or cold? (see page 123.) Did an initial lunar magnetic field decay away, as the earth's field appears to be doing (pp. 59-61), leaving the observed lunar remanent rock magnetism behind? What is the source of detected mass concentrations (*mascons*) buried beneath the lunar maria?

It is likely that radiocarbon dating will one day be applied to solar wind or meteorite-derived carbon in lunar samples. Even though the oldest absolute terrestrial dating scale is less than 10,000 years old (Bristlecone pines), the C-14 method is regularly extrapolated back 50,000 years.[28] Recently the promise was made for 100,000-year C-14 dating, using mass spectroscopic counting of actual carbon atoms rather than the usual monitoring of the rate at which C-14 decay is taking place.[29] This will be an extrapolation through 17 half-lives of carbon-14. Eventually the method will probably be credited with million- and billion-year ages, as stray C-14 atoms are detected and are fallaciously credited with being the remnant of ancient carbon concentrations.

The final conclusion must be that a complete understanding of the physical nature and history of the incredibly complex earth and moon, based on the scientific method, is entirely impossible. Similar to the dilemma arising from the precarious and conflicting lunar origin theories, the consideration of lunar geology leads one ultimately back to the Genesis account of a moon created by God. The moon, geologically fascinating even in view of its inhospitable condition, was created with the correct geological properties to fulfill its divine purposes of illumination and time-keeping.

---

[28]D. Wonderly, *God's Time-Record in Ancient Sediments* (Flint, Michigan: Crystal Press, 1977), p. 221. In this study, Daniel Wonderly interprets fossilized coral reefs as indicating a long earth history. A different survey of 76 various natural processes gives calculated earth ages varying all the way from 100 years to one-half billion years, the variability reflecting uniformitarian assumptions. Cf. H.M. Morris, "The Young Earth," *ICR Impact Series*, No. 17 (Institute for Creation Research, 2716 Madison Ave., San Diego, California 92116).

[29]D. Thomsen, "Radioisotope Dating with Accelerators," *Science News*, 113:2. (January 14, 1978), 29-30.

CHAPTER V

Transient
Lunar Phenomena

"Criticism of rationalistic pretensions will, of course, cut most deeply when it turns out that physical reality is not wholly conformable to what we, in a certain epoch, considered to be rational, and when the non-rationality, or even absurdity, of reality (or a part of reality) has to be acknowledged."

R. Hooykaas
(Religion and the Rise of Modern Science, *p. 44*)

The purpose of this chapter is to show that the moon is by no means the cold, dead body that cosmic evolutionists have traditionally held it to be. In sharp contrast to the view that the moon "died" some three billion years ago, there is much well-documented evidence of *present* activity on the moon's surface, and this may be just a small indication of what is taking place on the inside. Although most people are probably not aware of it, there have been numerous sightings of brief changes on the moon's surface. These include localized color changes, spots or streaks of light, clouds, hazes, veils, and other observations that speak of geological activity on the moon.

HISTORICAL BACKGROUND

In the early days of telescopic astronomy such events were

freely reported. Since there had not yet been built up a set of preconceived notions as to what could or could not happen on the moon's surface, men could publish their findings without fear of ridicule from the scientific community.

As the conception of the moon as a cold, dead body became more firmly entrenched in the centers of higher learning, it became increasingly less advisable to report observations of such events. One's reputation could be sullied, and there was usually no way that an observer could defend himself against accusations of slipshod technique or faulty interpretation of what he had seen. Because the transient phenomena are usually short-lived, the element of reproducibility, which is so desirable in science, was missing. Unless great haste could be exercised in notifying a second observer, a sighting would more often than not go unconfirmed, and anything that does not enjoy repeatability in science is invariably relegated to the "unwanted" category.

Thus it was that during the first half of the present century such reports were looked upon with disdain by the scientific establishment. And, since the transient phenomena could not be reconciled with the conventional theories of the day, it seemed expedient either to ignore them completely or to impugn their credibility. If the general reader were to learn of this side of lunar science, he had to do so not from the standard textbooks of astronomy, but from the "borderlands of science" writers. Charles Fort, for example, who specialized in documenting occurrences that science was unable to explain, wrote of transient lunar phenomena in 1923.[1] With great zest Fort described mysterious glows and points of light, colored shadows, and moving clouds on the lunar surface. For this and other reasons most people dismissed him as a crackpot.

A generation later Frank Edwards took up the cause. In 1959 he described several puzzling sightings in Mare Crisium and Aristarchus.[2] In 1964 he recorded an equally baffling

---

[1]Charles Fort, *New Lands* (New York: Ace Books, Inc., 1923).

[2]Frank Edwards, *Stranger than Science* (New York: Bantam Books, Inc., 1959).

event that took place in the Fra Mauro area.[3] To a great extent, however, the scientific world looked upon Edwards as a sensationalist who failed to check his facts.

Yet to those who studied the question carefully, the evidence for transient phenomena that had accumulated by the mid-sixties seemed overwhelming. No longer could publications on the moon ignore this aspect of selenography if they were to cover their subject properly. Ralph Baldwin, in *A Fundamental Survey of the Moon* (1965), devoted an entire chapter to "Changes on the Moon."[4] Willy Ley's classic *Ranger to the Moon*, which appeared the same year, included a more sizable chapter concentrating on just one type of event—the "red spots."[5]

Patrick Moore and Peter Cattermole devoted a substantial section of *The Craters of the Moon* (1967) to the subject. These writers observed in 1967 that:

> There has been a marked official change of view during the past decade. Before 1958, it was considered by professional astronomers that the moon must be totally inert—as had been claimed so long before by Beer and Mädler (1838). . . . In November 1958, however, N. A. Kozyrev, at the Crimea, observed activity in the crater Alphonsus. . . . This caused a definite switch of opinion. . . . And when further observations of short-lived colour patches were made at Flagstaff, in 1963, the transformation was complete. Since then, the Moon-Blink and other programmes have brought in further reports of transitory phenomena, and the reality of such 'lunar events' can hardly be questioned.[6]

In 1968 NASA published its *Chronological Catalog of Reported Lunar Events* (NASA TR R-277), which forms the basis for much of the factual content of this present discussion.[7] The *Catalog* is a carefully documented listing spanning

---

[3] Frank Edwards, *Strange World* (New York: Ace Books, Inc., 1964).

[4] Ralph Baldwin, *A Fundamental Survey of the Moon* (New York: McGraw-Hill Book Company, Inc., 1965).

[5] Willy Ley, *Ranger to the Moon* (The New American Library, New York, 1965).

[6] P. Moore and P. J. Cattermole, *The Craters of the Moon* (New York: W. W. Norton & Company, Inc., 1967), p. 75.

[7] B. Middlehurst, J. Burley, P. Moore, and B. Welther, *Chronological Catalog of*

the period from A.D. 1540 to 1967, giving in each case the date and time, the feature or location studied, a description of the event, the name or names of the observer(s), and the reference from which the information was taken. The *Catalog* contains 579 separate entries and is documented by a total of 250 references.

It has now even become acceptable to include a brief mention of transient phenomena in some of the more complete general astronomy textbooks.[8] A good summary of the *present* view of the situation is given in Nicholas Short's *Planetary Geology.* He states that

> over twelve hundred sightings of transients have been reported from more than a hundred locations; of these more than three hundred are associated with Aristarchus, seventy-five with Plato, and twenty-five with Alphonsus. . . . *These lunar transients may be signs of gas discharges or of current volcanic activity*[9] *(Italics added).*

Dr. Short includes a map of the moon's near side showing 118 locations where transient phenomena have been sighted—46 in the first quadrant, 28 in the second, 31 in the third, and 13 in the fourth (Fig. V-1).

What follows is a summary of the reports that have been published, based primarily on the listings in the *Catalog* (occasional supplemental material from other sources has been introduced where available). The reports are arranged chronologically within each topographical category; the number in parentheses given after each report is the official NASA catalog number. The events generally cover an area of no more than a few kilometers in dimension and extend over only a few hours in time.

The list of observations is long, yet one still questions its

---

Reported Lunar Events (NASA TR R-277 [1968], available from Clearinghouse for Federal Scientific and Technical Information, Springfield, Va. 22151).

[8]See George Abell, *Exploration of the Universe* 2nd edition (New York: Holt, Rinehart, and Winston, Inc., 1969), and D. Menzel, F. Whipple, and G. de Vaucouleurs, *Survey of the Universe* (Englewood Cliffs, N.J.: Prentice-Hall, Inc., 1970).

[9]N. M. Short, *Planetary Geology* (Englewood Cliffs, N.J.: Prentice-Hall, Inc., 1975), p. 64.

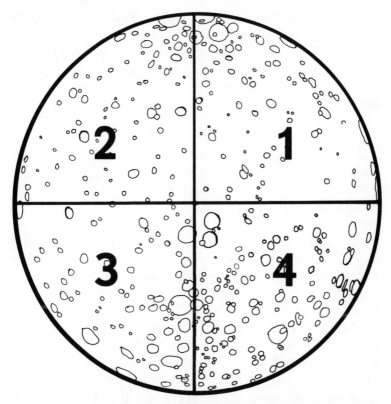

Figure V-1.   The moon's near side is divided into quadrants, or quarters as shown. The quadrant numbers follow the standard notation used for Cartesian coordinates. This is viewed as the moon appears to the naked eye. Note that a telescopic view of the moon is inverted from this figure unless an image corrector is used.

completeness. There is a definite element of serendipity involved in the observation of transient lunar events. Unless a competent observer happens to be looking in the right place at the right time using an instrument with adequate capabilities, the event will be missed. Then, the observer must be willing to report his sighting. One can only guess how many observers have been afraid to publish their findings for fear of ridicule. It is likely that down through the centuries *most* of the events have actually gone either unobserved or unreported, and that the list we have is but a mere sampling of transient activity on the moon's surface. Also, it is quite possible that the moon's *far* side has *its* share of transient phenomena. Herein lies a whole new area for investigation. The far side is, of course, inaccessible to earth-based telescopes, and it is questionable whether satellite photographs will prove to be a satisfactory methodology for a serious study of the subject in question.

THE MOON'S DARK SIDE

A certain percentage of the reports of transient phenomena come from the moon's *dark* side.[10] It is generally more difficult to pinpoint the exact geographical location of a dark-side event, so we are sometimes in doubt as to whether a given report concerns a mountain, a crater, or some other topographical feature. As noted in the NASA *Catalog*, there were more dark-side observations in the early days of optical astronomy when the magnifying power of the telescopes was smaller; hence the field of observation was larger, often including the whole moon.

The earliest listings in the *Catalog* actually antedate the invention of the telescope! These concern starlike appearances on the dark side, one in 1540 located in the region of the crater Calippus (Catalog No. 1), the other in 1587 directly between the points of the horns of a five- or six-day crescent (No. 2). In 1668 a bright starlike point was seen on the dark side by several New Englanders (No. 4).

Perhaps the most famous dark-side observations were those made by William Herschel, the discoverer of Uranus. His first sighting took place on May 4, 1783 (No. 20). Concerning this he wrote in a letter to J. H. de Magellan, a Portuguese scientist:

> I perceived in the dark part of the moon a luminous spot. It had the appearance of a red star of about the 4th magnitude.[11] It was situated in the place of *Hevelii Mons Porphyrites* (Archimedes), the instrument with which I saw it was a 10 feet [sic] Newtonian

---

[10]The moon's dark side here refers to the portion facing earth which is shadowed during non-full moon phases.

[11]Magnitude is the astronomical term for an object's brightness as seen from earth. The larger the magnitude number, the dimmer the object is. Most of the stars of the Big Dipper are of the second magnitude; a sixth magnitude star can just barely be seen with the naked eye.

Reflector of 9 inches aperture.[12]

Four years later, in March 1787, Herschel saw *three* bright spots in the same vicinity (No. 24). On April 19 of the same year he again saw three lights (No. 25). This time he referred to them as "three volcanoes," and made a formal report of his sighting to the Royal Society in the following words:

> I perceive three volcanoes in different places of the dark part of the new moon. Two of them are either already extinct, or otherwise in a state of going to break out ... The third shows an actual eruption of fire, or luminous matter.[13]

The following night he made still another observation of the Aristarchus area (No. 26), concerning which he wrote the following:

> The volcano burns with greater violence than last night. I believe its diameter cannot be less than 3" [three seconds of arc], by comparing it with that of the Georgian planet [Uranus]; as Jupiter was near at hand, I turned the telescope to his third satellite, and estimated the diameter of the burning part of the volcano to be equal to at least twice that of the satellite. Hence we may compute that the shining or burning matter must be about three miles in diameter. It is of irregular round figure, and very sharply defined on the edges. ... The appearance of what I have called the actual fire or eruption of a volcano, exactly resembled a small piece of burning charcoal, when it is covered by a very thin coat of white ashes, which frequently adhere to it when it has been some time ignited; and it had a degree of brightness, about as strong as that with which such a coal would be seen to glow in faint daylight.[14]

It is tempting to speculate that Herschel was looking at the same three points that were observed by Greenacre, Barr, and others in recent years.

Other reports of spots or points of light on the dark side came from Schröter in 1788 (No. 30), Schröter in 1792 (No. 57), W. Wilkins and Stretton in 1794 (No. 58), Piazzi in 1799 (No. 61), Gruithuisen in 1821 (No. 66), Fallows later in 1821 (No. 67), Gruithuisen in 1824 (No. 74), Rankin and Cheva-

---

[12]W. Ley, *Ranger to the Moon*, p. 71.

[13]*Ibid.*, p. 71.

[14]*Ibid.*, p. 71, from Herschel's *Collected Works*, 1912 edition.

lier in 1847 (No. 91), Hodgson in 1866 (No. 107), W. O. Williams in 1867 (No. 113), Valier in 1912 (No. 217), and Woodward in 1948 (No. 284).

THE MOUNTAINS

If indeed there are active volcanoes on the moon as Herschel maintained, the logical place to look for them would be in the mountainous areas. Several intriguing observations are listed in the literature, which shall be considered in the present section.

Schröter in 1789 observed a small point of light next to Mont Blanc in the lunar Alps. The light, which resembled a fifth magnitude star, was visible for 15 minutes (No. 50). In 1843 Gerling, while studying the lunar topography along the terminator (i.e., the line separating the lighted and dark portions of the moon), saw an unusually bright spot that glowed like a star. The spot was situated on a mountain peak just south of the lunar Alps (No. 89). Holden in 1888 reported a point of light on the southern edge of the lunar Alps when the mountains were in darkness. He stated that the brightness of the glow was approximately equivalent to a first magnitude star (No. 166).

Pico and Piton, two prominent bright peaks to the south and west of the lunar Alps, also figure in the present discussion. Schmidt reported a bluish shimmering patch of light near Pico in 1844 (No. 90). Pickering saw a haze spreading from the area in 1912 (No. 219), and Rawstron submitted similar reports for September 1 and October 1 of the year 1933 (No. 238, No. 239). Moore in 1958 watched as Piton became enveloped in an obscuring cloud-like mist (No. 407). The size of the cloud must have been considerable, for the mountain rises to a height of more than a mile above the surrounding plain (Mare Imbrium), and is not steep-sided. Schneller saw a red obscuration of the mountain in December 1960 and January 1961, but less intense than the first sighting (No. 426, No. 427). It may well be significant that both Pico and Piton appear to be volcanic peaks, complete with craters in their summits.

Reports of transient activity in the Teneriffe Mountains, located somewhat to the southwest of the crater Plato, make especially fascinating reading. Hodgson, in 1847, saw an intermittently glowing bright spot about one-fourth the angular diameter of Saturn when the mountains were in darkness (No. 92). Seven years later, in 1854, the veteran observer Hart watched two luminous fiery spots in the mountains for five hours. Interestingly, this sighting was made when the mountains were on the moon's *bright* side (No. 95). Hart described the event as

> ... an appearance I had never seen before on the surface of the moon though I have observed her often these last forty years... It appeared to me from the brightness of the light and the contrast of the color *to be two active volcanoes or two mouths of one in action.* [15]

In 1797 Caroché believed he saw "a volcano on the moon" in the vicinity of the Heraclides Promontory, part of the Jura Mountain Range (No. 59). Reports that the Promontory appeared blurred or misty came from Moore and Docherty in 1948 (No. 283, No. 287). In 1912 Franks perceived a "glowing line of light" extending through the Leibniz Mountains into the moon's dark side (No. 218). McCorkle saw a bright flare on the moon's dark side near the Carpathians in 1955. The flare was similar to a second magnitude star in appearance and lasted about 35 seconds (No. 347). Also in 1955 Lambert observed two flashes from the edge of the Taurus Mountains (No. 352). It should be noted that most of the reports from mountainous regions seem to come from in and around Mare Imbrium.

There clearly does *appear* to be volcanic activity on the moon. Those who study the facts carefully are finding this increasingly difficult to deny. Moore and Cattermole offer the following assessment of the situation:

> It seems therefore, that although major outbreaks on the moon belong to the remote past, traces of volcanic activity linger on there.... There is no evidence of major structural activity now in progress on the lunar surface, but there are minor transient phenomena which indicate mild volcanic activity." [16]

---

[15] B. M. Middlehurst, *et al., Chronological Catalog of Reported Lunar Events* (National Technical Information Services, 1968), p. 11.

[16] Moore and Cattermole, *The Craters of the Moon*, pp. 87, 152.

The reader should note carefully, however, that this is *not* tantamount to saying that there are full-scale volcanoes on the moon. Rather, we may well be dealing only with outgassings and expulsions of dust or ashlike materials.

THE CRATERS

The lunar craters are well represented in the reports of transient activity. While there are dozens of craters in which such observations have been made, there are *three* major sites that stand out above the others: Aristarchus, Plato, and Alphonsus, in that order.

The Crater Aristarchus

As stated earlier, there have been more than 300 reports connected with the Aristarchus area. Historically the sightings date back to William Herschel, whose observations have already been described under "Dark Side" phenomena. In subsequent years a great many observers reported glowing red spots or volcano-like events either in the crater itself or close to nearby Schröter's Valley. It is noteworthy that the Aristarchus sightings have remained consistently numerous from Herschel's time to the present.

On October 30, 1963, the American observers Greenacre and Barr saw three red spots—one on either side of Schröter's Valley, and one inside the rim of the crater Aristarchus. None of the spots persisted for more than 20 minutes, and the entire display was over in about 45 minutes (No. 440). On the same night Budine and Farrell observed that the Cobrahead (the enlarged termination of Schröter's Valley) glowed for a period of about 7 minutes (No. 441). The spot in Aristarchus was seen again on November 28, 1963, by Greenacre and others at the Lowell Observatory (No. 444). On this occasion a blue-violet haze was also seen. The same night Clyde Tombaugh, the discoverer of the planet Pluto, observed a pink spot on the west side of the Cobrahead for 35 minutes (No. 445).

On December 28, 1963, a red spot was seen in the crater Aristarchus by nine students at Hiroshima, Japan (No. 447). This time the red patch spread outward toward the nearby crater Herodotus! Unlike many of the other observations,

Kepler

Aristarchus

Oceanus
Procellarium

Mare
Imbrium

Copernicus

Alps

Apennines

Plato

Mare
Serenitatis

Mare
Frigoris

Mare Crisium

Mare
Foecunditatis

N

S

Grimaldi

Mare
Humorum

Gassendi

Tycho | Ptolemaeus

Mare Nubium

Mare
Nectaris

Mare
Tranquillitatis

Langrenus

Vendelinus

Figure V-2.   Major features of the moon (at full phase). In a telescope,
the moon (and everything else) appears inverted (that is, South is at the
top). However, by convention, the moon is now shown as it appears to
the naked eye with North at the top. (Photograph courtesy of the
Benjamin/Cummings Publishing Co., copyright 1972, from Astronomy
One, Hynek and Apfel.)

this was seen on the bright side of the moon only one day before full moon. Glows, flashes, and hazes have continued to be seen off and on in the Aristarchus-Herodotus-Schröter's Valley region up until quite recently. If one were to select an area of the moon to watch for transient events, the odds would be more favorable in this vicinity than any other.

It had been hoped that something more could be learned of these intriguing events through manned lunar exploration. The Apollo 19 mission, one of the three that was canceled for budgetary reasons, had been scheduled for a landing in Schröter's Valley and a descent into Aristarchus to find out the cause of the "red spots."[17] But one can only speculate what the results might have been. The question will now remain unresolved for some time to come.

The Crater Plato

Strange happenings have been described in and around the crater Plato for almost three centuries. Bianchini observed a reddish streak on the crater's floor during a lunar eclipse in 1685 (No. 9). Again in 1725 Bianchini saw a track of ruddy light like a beam crossing the part of the crater that was in darkness (No. 12). A yellow streak of light was seen in the darkened crater by Short, Stephens, and Harris in 1751 (No. 14).

In 1788, observers in Mannheim reported a bright spot on the dark side of the moon near Plato (No. 29). Eleven months later Schröter described a bright area in the crater resembling a thin white cloud (No. 42). Gruithuisen reported a bright fleck in the southeastern part of the crater on the night of December 8, 1824 (No. 75).[18] Four months later he noticed that the western part of the crater was brighter than

---

[17]"What's Next in Space," *U.S. News and World Report* 67:5 (August 4, 1969), 29.

[18]Directions as established by the International Astronomical Union in 1961 are used here. At that time, the classical, or astronomical, "east" was changed to the astronautical "west," which is in agreement with ordinary terrestrial mapping with east at the right and west at the left. Thus, as one looks at a full moon in the southern sky from the northern hemisphere he sees the eastern part of the moon on his right, the western part on his left.

the eastern part (No. 76).

From this point forward chronologically the reports become too numerous to describe individually. Dozens of reports describe bright spots in or near the crater, light streaks across Plato's floor and wall, and clouds or obscurations within the crater. The unusual appearances have continued during recent years.

Possible evidence of *permanent* changes in Plato comes from the very careful observer T. W. Webb. The detailed map of the lunar surface by Beer and Mädler in 1838, the best selenography available for many years, showed four light-colored striations across the floor of the crater, roughly parallel to one another.[19] Webb states that he found this configuration to be "greatly changed" by 1855, and still further changed in 1859 when he published his famous *Celestial Objects for Common Telescopes.*[20]

The Crater Alphonsus

Located near the center of the moon's near side, the crater Alphonsus has been the subject of several intriguing observations. This site is mentioned in several places in the NASA *Catalog,* beginning with the early part of the present century. Alphonsus has been described at times as veiled or indistinct, due apparently to some kind of outgassing phenomenon.

The first observation listed in the *Catalog* is a sighting by Flammarion in 1906; however, no details are given concerning this event (No. 212). In 1937 Dinsmore Alter reported a milky appearance on the crater's floor (No. 246). On October 26, 1956, the same observer made a number of photographs of Alphonsus in both infrared and ultraviolet light. The pictures taken in the shorter wavelength were less distinct, indicating the presence of a haze (No. 381).

So convincing were these results that the Russian astronomer Kozyrev began studying the crater spectroscopically in an effort to ascertain the nature of this gas should another

---

[19] T. W. Webb, *Celestial Objects for Common Telescopes* (New York: Dover Publications, Inc., 1962 Edition), I, p. 126.

[20] R. B. Baldwin, *A Fundamental Survey of the Moon* (New York: McGraw-Hill, 1965), p. 111.

such event occur. His efforts bore fruit on November 3, 1958, when an effusion of gas occurred over the central peak of the crater (No. 409). As this occurred, the image of the peak became, in the words of Kozyrev himself, "strongly washed out and of an unusual reddish hue."[21] Two hours later he noted that the central peak glowed with unusually intense whiteness. But as he watched, its brightness fell back to normal. The next night nothing unusual was observed.

At first Kozyrev was tempted to dismiss the event as simply a "change in the quality of the observing conditions," but the spectrograms he had obtained testified otherwise. Two spectrograms had been made during the event, and a third after the crater had returned to normal. The first showed a distinct attenuation in the violet portion of the spectrum. Kozyrev attributed this to an outpouring of a reddish dust or ash. The second spectrogram was most remarkable: it showed broad emission bands characteristic of the element carbon in its diatomic molecular form, C_2. Presumably they were caused by an efflux of a carbon-containing gas after the dust or ash had been ejected. The exact chemical nature of the gas has become a subject of lively debate among astronomers. The third spectrogram showed nothing out of the ordinary, and later photographs of the area demonstrated that no permanent change had taken place in the crater.[22]

Here, then, was a definite milestone in the observation of transient lunar phenomena. An actual spectrogram had been made giving clues to the identity of the substance being ejected, and the spectrum was seen to return to normal after the event had terminated. Alphonsus was watched more closely from this time on. Numerous observers have since reported red glows and flashes within the crater (especially near the central peak), and several have obtained spectra with absorption lines. The Alphonsus events listed in the NASA

---

[21]For a good discussion of Kozyrev's work including plates of the actual spectrograms see: D. Alter, *Pictorial Guide to the Moon* (New York: Thomas Y. Crowell Co., 1967), pp. 147-49.

[22]NASA *Catalog* numbers 410, 412, 413, 414, 417, 423, 424, 503, 544, 549, 560, 561, 574.

Catalog were noteworthy for their frequency of occurrence.
Many of these are only a few days or weeks apart.[23]

THE MARIA

A few of the dark lowland regions of the moon have also
been mentioned in the reports. Eight of these—Mare Crisium,
Mare Humorum, Mare Imbrium, Mare Nectaris, Mare
Nubium, Mare Serenitatis, Mare Tranquillitatis, and Mare
Vaporum—are listed in the *Catalog.*

Mare Crisium

Mare Crisium, a dark oval lowland region in the moon's
first quadrant, has occasionally been observed to exhibit a
hazy or cloudy appearance, and, more rarely, unexplained
lighting effects. Cassini reported a nebulous appearance in
1672 (No. 7). Eysenhard observed four bright spots in the
mare in 1774, and described a peculiar behavior of the terminator, which happened to be located in this region at the
time of his observation (No. 16). The terminator exhibited a
slow reciprocating motion over Mare Crisium, yet over Mare
Fecunditatis the terminator appeared perfectly normal.[24]

A small nebulous bright spot was seen on the northern
edge of Mare Crisium by Schröter in 1788 (No. 38). A black
moving haze or cloud was seen for two days by Emmett in
1826 (No. 79, No. 80). On the second day the cloud was
noticeably less intense than the first day. In 1865 Ingall observed a point of light like a star and a misty cloud (No. 102).
Occasionally a "bright spot" or "white patch" has been seen
east of Picard, a crater located within the mare (No. 128, No.
133, No. 139). On May 19, 1882, J. G. Jackson observed a
cloud just east of Mare Crisium against the Agarum Promon-

---

[23]Other craters in which transient phenomena have been observed are
Agrippa, Alpetragius, Anaximander, Archimedes, Arzachel, Atlas, Bessel, Byrgius,
Callipus, Carlini, Clavius, Copernicus, Dawes, Eratosthenes, Eudoxus, Gassendi,
Godin, Grimaldi, Helicon, Humboldt, Kant, Kepler, Kunowsky, Littrow, Macro-
bius, Manilius, Marius, Mersenius, Messier, Peirce, Picard, Pitatus, Plinius, Posi-
donius, Proclus, Ptolemaeus, Riccioli, Ross, Schickard, Theophilus, Timocharis,
Triesnecker, and Tycho.

[24]See Webb, *Celestial Objects,* pp. 106-07, for a more complete description of
this remarkable observation.

tory. The cloud measured at least 100 miles by 40 or 50 miles (No. 150)! No trace of the cloud could be detected the following day. The same observer submitted similar reports for July of 1882 and May of 1883 (No. 151, No. 155). In 1915 Thomas observed a particularly bright spot like a star on the northern edge of the mare (No. 225).

On two separate occasions in 1927, H. P. Wilkins noted that the crater Peirce A in Mare Crisium was completely obscured (No. 231, No. 232). Normally, the crater is clearly discernible. On July 21-22, 1948, Patrick Moore observed that the floor of the crater became almost featureless for several hours (No. 282). Later the same summer Moore watched as two areas east of Picard became featureless, again for several hours (No. 285).

The observations in and around Mare Crisium speak of outpourings of gases or vapors, and/or finely divided particulate materials. Yet the actual openings (fissures?) from which they issue have not yet been located. The unusual lighting effects raise some perplexing questions. Clearly, much remains to be learned about this fascinating area.

Mare Vaporum and Mare Nubium

Mare Vaporum, just north of center on the moon's near side, is another region that is well known to lunar observers. Schröter and Olbers in 1797 reported vapors emanating from the mare (No. 60). Mare Nubium contains an unusual formation called the Straight Wall, thought to be a fault scarp. The scarp is some 60 miles long and 1,200 feet high. Its face makes an angle of 41 degrees with the horizontal. A report by Capen in 1955 concerned an obscuration of the small craters on the west side of the Straight Wall, while craterlets on the east side of the wall were observed with normal clarity. In 1824 Gruithuisen saw a bright area in the mare which he estimated to be 100 by 20 kilometers (No. 74). The mare was on the moon's dark side at the time of the observation. More recently Sanduleak and Stock, using photoelectric photometry, detected a strong anomalous enhancement of radiation in this region during a lunar eclipse in 1964 (No. 510). It is an intriguing fact that the crater Bullialdus, a frequently reported moonquake epicenter (i.e., an area on the surface directly above a moonquake source), is located in this same region.[25] Perhaps there is a connection.

SOME PROPOSED MECHANISMS

What causes the transient phenomena? It is quite possible that we are dealing with more than one type of phenomenon in this study. If so, multiple mechanisms may be involved, and the problem could be considerably more complex than it first appears. Some of the possible mechanisms that have been proposed will be described and evaluated.

Herschel's volcano hypothesis has been notorious for its lack of acceptance among astronomers generally. Webb, writing on the crater Aristarchus in his *Celestial Objects*, stated, "One of Herschel's very few errors was his taking it for a volcano in eruption."[26] Garrett P. Serviss, a science writer of bygone days, stated,

> It remained almost the only serious blot upon Herschel's record as an observer. He had described the appearance of a supposed eruption too carefully to admit any question as to his meaning. And yet, it seemed, a mere tyro in astronomical observation could hardly be deceived in such a manner, much less the most famous astronomer of his time."[27]

While we may tend to sympathize with the venerable Herschel, there *are* some serious problems connected with the volcano hypothesis. For one thing, the glowing areas observed seem to be too extensive for volcanoes. According to Herschel's own report, the active area whose size he estimated by comparison with Uranus and Jupiter's third moon, was about 3 miles in diameter. Other observers have reported glowing areas up to 10 or more miles in diameter. It is highly questionable whether the actual glowing portion of any volcanic eruption could extend over this great an area. Another difficulty is that no evidence of fresh lava flows has ever been detected in such regions, even on the most detailed photo-

---

[25] J. Ashbrook, "Astronomy," *1973 Britannica Yearbook of Science and the Future* (1973), 184-85.

[26] Webb, *Celestial Objects*, p. 129.

[27] Garrett P. Serviss, "New Light on a Lunar Mystery," *Popular Science Monthly* 34 (December, 1888), 158-61.

graphic plates.

Because the idea of any type of volcanic activity on a "dead planet" was repulsive to most astronomers, it became fashionable to explain the phenomena by special lighting effects such as a low sun angle. Prof. Holden, a director of Lick Observatory almost a century ago, claimed that Herschel's observations were in fact attributable to "specially brilliant and favorable illumination."[28] Holden believed that he himself had duplicated Herschel's observations. However, it should be noted that Holden's sightings were made in the lunar Alps rather than the Aristarchus region. A mountainous topography will obviously give rise to different lighting effects from a topography consisting of craters and valleys. Also, the solar illumination hypothesis fails to explain the dark-side observations, and these are, as has been noted, quite numerous in the literature. A great deal of care was exercised in compiling the NASA *Catalog* to eliminate reports attributable to unusual lighting conditions. These effects included "earthshine (strongest during the first and last three days of a lunation, [i.e., the interval between two returns of the new moon]), sunshine on peaks just beyond the terminator, differences in albedo [i.e., the percentage of light reflected from an object] and color in small regions, and multiple reflections from crater walls."[29] Several instances of this type are cited and described, but omitted from the general listing.

Now, of course, there is more complete information on which to base hypotheses than in former years. It is known today, for example, as a result of Kozyrev's work, that outgassing (i.e., ejection of a gas or vapor from within the moon) plays a part in at least some of the events. Middlehurst and Moore, two of the compilers of the NASA *Catalog,* report that the locations of the events fall into three classes:[30] 1) sites peripheral to the maria; 2) ray craters; 3) ring plains with dark or partially dark floors. They note that there have been no authenticated reports from the rugged highland

---

[28]*Ibid.*, 159.

[29]Middlehurst, *et al.*, *Chronological Catalog*, p. 3.

[30]B. M. Middlehurst and P. A. Moore, "Lunar Transient Phenomena: Topographical Distribution," *Science*, 155 (January 27, 1967), 449-51.

region in the southeastern part of the moon. Thus the total number of reports from the fourth quadrant is significantly smaller than it is for the other three quadrants. Middlehurst and Moore believe the aforementioned distribution to be excellent evidence for tying the transient phenomena to volcanic activity in the moon's interior. They also state in the same paper the very remarkable fact that the most probable time for such events to occur is within two days of perigee (i.e., the point in the moon's orbit which is closest to earth). This is the time when tidal stresses are greatest, and presumably most effective in dislodging trapped gases. Also, it is now known that tidal stresses can trigger moonquakes, which in turn may release trapped gases by mechanical vibration.

It is thus possible that the transient occurrences are energized by heat and pressure from the lunar interior. If so, these phenomena are closely related to the question of the moon's internal heat. It has been exciting to watch the modern-day revolution in thinking on the latter question. There was a time in the not-too-distant past when the theorists could say, "The moon *must* have a cold interior because it has had several billion years to cool." It is now, of course, understood that the moon is *not* cold inside. Speaking of the interior, Baldwin writes, "The moon is now hot and always has been hot. It is close to the melting point at each level beneath a cool outer zone."[31] He believes that temperatures on the order of 1000°C to 1500°C are not unwarranted by the data. This represents a complete turnabout in majority opinion, and it has taken place very rapidly. As recently as 1970, some scientists still considered the moon to be cool or cold inside.[32] The hot lunar interior view was greatly strengthened by the findings of Apollo 15 and 17 thermal flow experiments.[33] Heat sensors placed at various depths showed a thermal gradient two to three times higher than had been expected. This means, in very simple terms, that there is

---

[31] R. B. Baldwin, "Summary of Arguments for a Hot Moon," *Science*, 170 (December 18, 1970), 1297.

[32] *Ibid.*, 1297-98.

[33] See N. M. Short, *Planetary Geology*, pp. 183-84.

considerably more heat inside than had been anticipated. An initial response to this has been to ascribe the heat to radioactivity. However, this approach has already been tried and found wanting for the earth. Leet and Judson, for example, explain:

> Resorting to radioactivity as the source of the earth's heat seems forced and unnatural in the light of relatively unlimited supplies of primeval heat that could have been left over from the earth's formation. Radioactivity was principally a means for trying to explain local pockets of magma, yet radioactive elements have *not* been extruded at volcanic vents in the ocean basins.[34]

Interestingly, these authors do not even claim to know what temperatures are called for in the mantle: "if we assume that mantle materials are not solid, they could be at an indefinitely high temperature."[35]

Possibly there is a continual buildup of gas within the moon's interior as a result of the chemical reactions associated with lunar vulcanism. From time to time the pressure developed by this gas would have to be released in various places through faults or fissures in the crust. The fact that obscurations have been observed near the Straight Wall, a structure that has all the identifying characteristics of a fault scarp, lends support to this view. In some places the gas, in order to escape, might have to break through a "plug" of ashlike material, causing, in the process, the expulsion of both particulate *and* gaseous materials. This accords well with Kozyrev's observation that there was first an ejection of a reddish dust or ash, followed later by the outgassing which gave rise to the much publicized spectrogram. Alternatively, in some cases, the observed dust might be blown from the surface as the escaping gases expand into the surrounding vacuum.

It remains yet to consider possible mechanisms for the anomalous glows and points of light that have been associated with so many of the events. One possibility might be a

---

[34] L. D. Leet and S. Judson, *Physical Geology*, 4th ed. (Englewood Cliffs, N.J.: Prentice-Hall, Inc., 1971), p. 84. Emphasis added. Cf. Florence J. Leet and L. Don Leet, "The Earth's Mantle," *Bull. Seis. Soc. Am.*, Vol. 55 (1965), 619-25.

[35] Leet and Judson, *Physical Geology*, p. 84.

fluorescent effect produced by the sun's ultraviolet rays. Whipple states,

> There is evidence for fluorescent light on the lunar surface arising possibly from solar ultraviolet light and high-energy particles in space. It shows best on sunrise areas of the moon where the solar light probably causes the surface layer to radiate the energy stored up by impacts with ions of the solar wind.[36]

William Corliss has noted that the enhanced lighting of the moon during a total lunar eclipse that occurred while a terrestrial aurora was in progress may serve to correlate some transient lunar phenomena with charged particles from the sun.[37]

A. A. Mills offers another explanation for the lights—electrostatic glow discharge:

> The release of trapped gases, and their rapid expansion, must lead to disturbance of the fine dust covering the lunar surface. Now it has long been known that movement and dispersion of finely divided material result in separation of charge and the generation of considerable electrostatic potentials. Faraday, and later Rudge and others, showed that a positive charge was left on the plate of an electroscope if sand was blown from it. I have found charges of the same sign to be produced by the dispersion of powdered samples of basalt, olivine, ilmenite, limonite, ash-flow deposits and chondritic meteorites. . . . The generation of considerable electrostatic potentials in fluidized beds has been reported. On a large scale, many industrial accidents have been traced to the movement of dust-laden air generating potential differences sufficient to promote an incendiary spark.[38]

Could this be the explanation for the many "flashes of light" on the moon's surface that have been reported in the literature? Mills adduces in further support of this view the fact that lightning displays in volcanic ash clouds are a well-known effect; they have been reported from the time of

---

[36] Fred Whipple, *Earth, Moon and Planets,* 3rd ed. (Cambridge, Mass.: Harvard University Press, 1968), p. 132.

[37] William R. Corliss, *Strange Universe: A Sourcebook of Curious Astronomical Observations* (Glen Arm, Md., 1975), p. A1-126. This eclipse was reported in the *Monthly Notices of the Royal Astronomical Society* by Forster in 1848.

[38] A. A. Mills, "Transient Lunar Phenomena and Electrostatic Glow Discharges," *Nature* 225 (March 7, 1970), 929.

Pliny to the present. Also he notes that terrestrial dust storms cause severe disturbances in the earth's electric field. Thus we have a fairly satisfying mechanism for the "flashes." What can we conclude about the "glows"?

> At reduced pressures charge equalization is facilitated by glow discharge. Experiments indicate this mechanism to be most important in the pressure range 10-0.1 torr. [One torr is the pressure due to a column of mercury one millimeter high]. Studies of the light accompanying the generation of frictional electricity *in vacuo* were begun by Hauksbee in 1704 and continued by others, but little has been added since.

> I therefore propose that the obscuration associated with TLP's (transient lunar phenomena) is due to fine dust raised by lunar degassing. In certain areas, sporadic escape of accumulated gas occurs through channels opened by tidal stress. Movement and separation of the heterogeneous particles result in separation of charge and buildup to promote glow discharge through the transient gas phase. The predominance of hydrogen in the discharge gives the reddish tint.[39]

So far as proposed mechanisms are concerned, this paper by Mills probably represents the most comprehensive analysis of the subject to date. Extending Mills' approach, one is tempted also to speculate that there may be a correlation between the dust storms on Mars and the flashes of light that have been reported on the Martian surface. Several observations of this nature are on record.[40]

CONCLUSION

Whatever the mechanism or mechanisms involved, the transient lunar phenomena are there. No longer can the reports all be ascribed to faulty observing techniques or erroneous interpretations. The transient events indicate that the moon is still active geologically. It is not the cold, dead body that the pre-space-age theories had depicted. Yet, it *should* be cold and dead if it is indeed billions of years old.

Someone has aptly noted that there are three stages in the acceptance of a new scientific idea:

---

[39] *Ibid.*

[40] Corliss, *Strange Universe*, pp. A1-139ff.

(1) It could not possibly be true.

(2) What difference would it make if it *were* true?

(3) We knew it long ago.

In the case of the transient lunar phenomena, the third stage is now hard upon us. Now that a probable source of energy for the transient events is recognized to exist (namely, a hot lunar interior), the events themselves are being legitimized. As unpalatable as this has been for many, ever-increasing numbers of serious-minded astronomers are coming to recognize their existence. After more than four centuries of observations, the transient lunar phenomena are finally being accepted, be it ever so grudgingly, into the fabric of conventional science.

The heavens beyond the moon show unending variety. (U.S. Naval Observatory)

The Milky Way open cluster Pleiades, mentioned in Job 9:9, Job 38:31, and Amos 5:8.

The globular star cluster in Hercules (M13), on the outer edge of the Milky Way. Thousands of gravitationally bound stars are visible.

A spiral galaxy (M51) in Canes Venatici, made up of 100 billion stars.

A time exposure of the Milky Way center. The track is that of a passing artificial satellite. (U.S. Naval Observatory)

Glaciers moving down the Alaskan Tanana River Valley. Satellite surveys are valuable for land use planning. (NASA)

Astronaut Schmitt (Apollo 17) stands next to a lunar boulder which has been split. The area is the Taurus-Littrow landing site. (NASA)

The photomosaic of the Martian satellite Phobos shows chaotic linear grooves, about 500 meters (0.3 miles) wide. They may be due to surface fracturing caused by tidal forces of Mars and impacts of large debris. The picture on the left is from raw imaging data; the picture on the right has been computer enhanced. (NASA)

The 240-megawatt tidal power plant on the Rance River in France. It is composed of a lock, a 163-meter (535 foot) dike, and a gate dam. Energy is provided by the tidal movement of water, caused by the moon's gravitational attraction. (French Embassy Press and Information Service)

CHAPTER VI

Lunar Distinctives

"For myself, faith begins with the realization that a supreme intelligence brought the universe into being and created man. It is not difficult for me to have this faith, for it is incontrovertible that where there is a plan there is intelligence—an orderly, unfolding universe testifies to the truth of the most majestic statement ever uttered—'In the beginning God.'"

> *Arthur H. Compton*
> *1923 Physics Nobel Prize Winner*
> *(From a speech published in the*
> Chicago Daily News, *April 12, 1936)*

INTRODUCTION

The hundreds of billions of stars in the Milky Way galaxy and the nine planet members of the solar system vary dramatically in their physical properties. In fact, the Bible declares that "There is one glory of the sun, and another glory of the moon, and another glory of the stars: for one star differeth from another star in glory" (1 Cor. 15:41).

Orion's right shoulder star Betelgeuse is a 560,000,000-kilometer-diameter variable red giant; but how different "in glory" is the compacted white dwarf companion of Sirius with a density of one million grams per cubic centimeter, about 3,000 times denser than anything on Earth. One matchbox-sized nugget of this white dwarf material, if returned to earth, would weigh several tons!

Within the solar system, the planets Saturn and Uranus spin within multiple ring halos, while Jupiter's southern hemisphere displays a swirling lens-shaped red spot three times larger than the earth. This amazingly rich variety of physical properties continues even down to the scale of planetary satellites. Jupiter's orange moon Io may experience a methane (CH_4) snowfall each time it passes into Jupiter's shadow.[1] The satellite Phobos completes three rapid revolutions around Mars for each rotation of the planet, thus rising two times in the west during each day-night cycle on Mars.

Although the earth has only one large natural satellite, the moon is by no means commonplace. The moon clearly shows God's creative handiwork by its beauty and usefulness to earth. It is no wonder that John Adams, the second president of the United States, based his belief in the existence of God on such things as "the amazing harmony of our solar system" and "the stupendous plan of operation."[2] The significance of one major feature of our lunar satellite, namely, its large size relative to the earth, will be considered in this chapter.

LUNAR BRIGHTNESS

The mass ratio of the moon-earth pair is more than 10 times greater than that for any other planet-satellite pair in the solar system. Except for the Triton-Neptune pair, the difference is 100-fold (Table VI-2). While several satellites are heavier than the moon, no other planet possesses a satellite having a mass which is such a substantial fraction of the

---

[1] R. O. Fimmel, W. Swindell, and E. Burgess, *Pioneer Odyssey-Encounter with a Giant* (Available from Scientific and Technical Information Division, NASA, Washington, D.C., 1974), p. 88. The methane snowfall idea has been challenged by a more recent study. U. Fink, H. P. Larson, and T. N. Gautier, "New Upper Limits for Atmospheric Constituents on Io," *Icarus*, 27 (1976), 439. J. Eberhart, "Cloud-Gazing at Io," *Science News*, 112, 20 (Nov. 12, 1977), 332-33. Nevertheless, Io remains a fascinating moon. Other unique properties include its modulation of Jupiter's decametric radiation, its surrounding cloud of sodium emission, and a surface that may be covered with salt, much like the salt flats in the American west.

[2] L. H. Butterfield, Ed., *The Adams Papers* (Cambridge, Mass.: Harvard University Press, 1961), I, pp. 27-30.

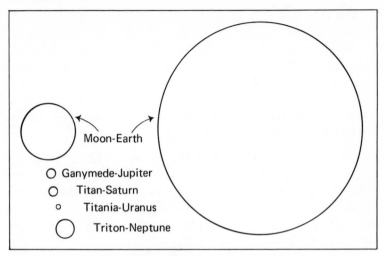

Figure VI-1. Sizes of larger satellites compared to their planets. The large circle represents the particular planet.

planet's mass (Fig. VI-1). For this reason the moon is often called a "secondary" or "double planet" companion to earth. Suppose the moon remains at its present position in order to maintain the monthly cycle of lunar phases, but its mass were reduced 100 times to make it an "average-sized" solar system satellite.[3] The moon's diameter would be decreased from its present 3,479 kilometers (km) to only 748 km (a 78.5% decrease[4]). Since the amount of light reflected from the moon depends on its area, proportional to the square of its diameter, its full moon brightness would be reduced to just 5% of

---

[3]The German astronomer Johannes Kepler (1571-1630) empirically made the observation that a satellite's orbital distance from its planet is independent of the satellite mass. Isaac Newton (1642-1727) later rederived Kepler's equations for satellite motion in a general way and showed that the satellite mass is a very small correction. See G. Abell, *Exploration of the Universe* (New York: Holt, Rinehart, and Winston, 1969), p. 81.

[4]With a diameter d and a mass density ρ, the moon's mass m is given by
$$m = \frac{\pi \rho d^3}{6}$$
If m is decreased 100 times, then d is decreased to d', where
$$d' = \left(\frac{1}{100}\right)^{1/3} d$$
$$= 0.215d$$
Taking d as 3,749 km, then d' is 748 km.

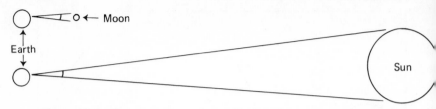

Figure VI-2. The sun and moon appear to be the same size in the sky. Diagram is not to scale.

its present value.[5] Even then the moon would be the dominating object of the night sky, many times brighter than the next rival, the brilliant planet Venus. Nevertheless, the moon's present illumination would be diminished 20 times by such a size decrease. Its large relative size is definitely necessary for it to provide significant evening light.

The Book of Genesis states: "And God made two great lights; the greater light to rule the day, *and the lesser light to rule the night:....*" (Gen. 1:16). The Hebrew term translated *light* (ōr) in this passage is flexible enough to include light reflectors such as the moon and the planets. The fact that the moon, which is a reflector, and the sun, which is far from being the largest star, are named "two great lights" is perfectly consistent with the language of appearance which the Bible uses throughout, as has already been explained (cf. pp. 66-67). A God who could not communicate with men in terms that they could understand would be limited indeed. From our human perspective, the only light that truly dominates the night is the moon.

SOLAR ECLIPSES

The combination of size and distance of the moon from

---

[5]The brightness B of a light source is proportional (symbolized by $\propto$) to the object's area, written in terms of its diameter as

$$B \propto \frac{\pi d^2}{4}$$

If the diameter is decreased to 0.215 of its actual size, the new brightness B' is

$$B' = (0.215)^2 B$$
$$= 0.046 B$$

Brightness in astronomy is described in terms of magnitude numbers, such that a first-magnitude star is 2.512 times brighter than a second-magnitude star. The result of decreasing the lunar mass 100 times would be a magnitude change between 3 and 4 units.

earth results in the special situation that the angular size of the sun and moon as seen from earth are nearly equal (Fig. VI-2). The moon's diameter is 400 times smaller than that of the sun but it is likewise 400 times closer to the earth. The visual angle subtended by the objects, measured in degrees, is found by the following calculations:

$$\text{Angular size of moon} = \frac{360°}{2\pi} \times \frac{\text{moon diameter}}{\text{earth-moon separation}}$$

$$\text{Angular size of sun} = \frac{360°}{2\pi} \times \frac{\text{sun diameter}}{\text{earth-sun separation}}$$

The results depend on the denominator distances used, which vary due to orbital eccentricity. Perigee and perihelion points are respectively the closest approaches of the moon and sun to the earth. Apogee and aphelion are the far points. The calculated range of angular sizes showing the overlap between the moon and sun is as follows:

Distance (kilometers)		Angle in degrees
Moon diameter	3,479	
Perigee	356,412	0.559
Apogee	406,686	0.490
Sun diameter	1,391,079	
Perihelion	147,224,000	0.541
Aphelion	156,955,000	0.508

Because of this similarity, the moon is able to eclipse the sun exactly when the respective orbit planes intersect and distances are matched. Computer studies show that this phenomenon is unique among the presently known moons of the solar system.[6] Other moons provide at best a partial, annular, or massively total eclipse of the sun for their planets. The size-distance balance between the moon and sun is usually labeled a "surprising coincidence" or a "lucky accident" in nature. However, this phenomenon actually points to yet another detail of design in the moon's creation, for man's bene-

---

[6]M. Mendillo, and R. Hart, "Resonances," *Physics Today*, 27:2 (February, 1974), 73. The number of discovered moons in the solar system is rapidly increasing, since this is an area of intense study. The present count is 35; in 1974 there were 32. Besides the earth-moon system, the only other case of an exactly total solar eclipse is the Neptune-Pluto planet pair.

fit and for God's glory.

Before the year 1930, when Bernard Lyot of France solved the problem of observing the corona without the aid of an eclipse,[7] the moon was essential to the discovery of the secrets of the sun.

> A few thousand miles above the photosphere, the solar atmosphere is so thinned and transparent that it becomes virtually invisible in the glare of light from the photosphere. But when the brilliant disk is masked by an eclipse, we discern a very interesting profile. Then the sun's corona, or extended halo, becomes briefly visible in one of the most dramatic of all nature's spectacles.[8]

It is quite impossible for many people to accept this vastly important "perfect fit" as accidental.

In his classic work *The Mysterious Numbers of the Hebrew Kings*,[9] Edwin R. Thiele has emphasized the crucial importance of eclipses for unlocking the otherwise hopeless confusion of ancient historical records. With regard to the various cuneiform copies of the Assyrian eponym canon, for example, Thiele points out that

> one item of unusual importance is a notice of an eclipse of the sun which took place in the month Simanu, in the eponymy of Bur-Sagale. Astronomical computation has fixed this as June 15, 763 B.C. With the year of the eponymy of Bur-Sagale fixed at 763 B.C., the year of every other name of the complete canon can likewise be fixed. The Assyrian lists extant today provide a reliable record of the annual limmu officials from 891 to 648 B.C., and for this period they provide reliable dates in Assyrian history.[10]

For ancient Near Eastern history following the middle of the eighth century B.C., the canon of Ptolemy is useful:

---

[7] H. Friedman, *The Amazing Universe* (Washington, D.C.: National Geographic Society, 1975), p. 59.

[8] *Ibid.*, p. 58.

[9] E. R. Thiele, *The Mysterious Numbers of the Hebrew Kings* (Chicago: University of Chicago Press, 1951); cf. Revised Edition (Grand Rapids: Wm. B. Eerdmans Pub. Co., 1965), and popular version, *A Chronology of the Hebrew Kings* (Grand Rapids: Zondervan Pub. House, 1977). See also R. A. Parker and W. H. Dubberstein, *Babylonian Chronology 626 B.C.-A.D. 75* (Providence, R.I.: Brown University Press, 1956).

[10] *Ibid.*, p. 44. Thiele provides the complete eponym list on pp. 287-92.

What makes the canon of such great importance to modern historians is the large amount of astronomical material recorded by Ptolemy in his *Almagest*, making possible checks as to its accuracy at almost every step from beginning to end. Over eighty solar, lunar, and planetary positions, with their dates, are recorded in the *Almagest* which have been verified by modern astronomers. The details concerning eclipses are given with such minuteness as to leave no question concerning the exact identification of the particular phenomenon referred to, and making possible the most positive verification.[11]

The solar eclipses referred to by Ptolemy are confirmed as March 19, 721 B.C.; July 16, 523 B.C.; and April 25, 491 B.C. Thiele concludes:

Since Ptolemy's canon gives precise and absolutely dependable data concerning the chronology of a period beginning with 747 B.C., and since the Assyrian eponym canon carries us down to 648 B.C., it will be seen that there is a century where these two important chronological guides overlap and where they may be used as a check upon each other.[12]

It is this unique capacity of the moon to eclipse the sun exactly, when the respective orbit planes intersect, that unlocks the exact chronology and thus the true history of a large segment of the first millennium B.C. The significance of the eclipse data for Biblical studies is incomparably great, for it provides confirmation, unavailable for well over two thousand years, that the chronological systems employed by Old Testament scribes were perfectly accurate.[13]

Despite the obvious importance of solar eclipses, not everyone accepts their significance. The futile arguments of the atheist display an ignorance regarding the value of eclipse

---

[11]*Ibid.*, p. 46; cf. p. 293 for Babylonian and Persian rulers according to the Canon of Ptolemy.

[12]*Ibid.*, p. 47. Otto Neugebauer concludes: "Ptolemy's famous statement that practically complete eclipse records from Babylon existed, beginning with the mid-eighth century B.C., is undoubtedly correct. Thus at an early period the material was available from which one could establish the periodicity of lunar eclipses." *A History of Ancient Mathematical Astronomy* (New York: Springer-Verlag, 1975), p. 549.

[13]See J. C. Whitcomb, Jr., *Chart of Old Testament Kings and Prophets* (Winona Lake, Ind.: BMH Books, 1977).

phenomena:

> Why, if its movements in space are guided by intelligence,
> should the moon eclipse the sun? Why should its orbital "adjust-
> ment" be so badly timed as to bring this satellite between the
> earth and its luminary and blot out the light? A total, or even a
> partial eclipse of the sun is a mere case of the moon "getting in
> the way."[14]

In actuality, an eclipse of the sun is universally recognized as
an awesome event. The authors of *Survey of the Universe*
fittingly expressed their eclipse experiences:

> A total eclipse of the sun is, without question, the most magnifi-
> cent of all astronomical phenomena. For about one hour after
> *first contact*, when the moon's edge first encroaches on the sun's
> disk, the moon slowly eats away the west side of the sun until
> finally, at *second contact*, it completely covers the sun's disk.
> During this interval, the sun's brightness diminishes, the air cools,
> and the sounds of animal life around slowly die away as the
> uncommon event unfolds. . . . The eerie quality suddenly intensi-
> fies a thousandfold as the solar corona flashes into view around
> the darkened sun. If the eye has been adapted to the diminishing
> light intensity by very dark glass, the corona registers as a brilliant
> spectacle even though it radiates no more light than a full moon.
> Totality can last for a maximum of 7½ minutes, but most eclipses
> are only several minutes' duration. The observer, who may have
> traveled many thousands of miles to see the eclipse, receives in
> those few minutes soul-satisfying reward for his efforts.[15]

THE MOON'S SHAPE

The relatively large size of the moon establishes its familiar
spherical shape. The lunar mass is actually not much larger
than the minimum necessary for such a solid body to assume
a spherical form. After performing the necessary calculations,
Kopal concludes:

> The pressure inside the moon should, therefore, be of the order
> of 10 kilobars throughout most of its mass. This is about 10 times
> the crushing strength of granite. . . . Given a sufficiently long time
> (no doubt short in comparison with the age of the moon),

---

[14]W. Teller, *The Atheism of Astronomy*, p. 94.

[15]D. H. Menzel, F. L. Whipple and G. de Vaucouleurs, *Survey of the Universe*
(Englewood Cliffs, N.J.: Prentice-Hall, Inc., 1970), p. 219.

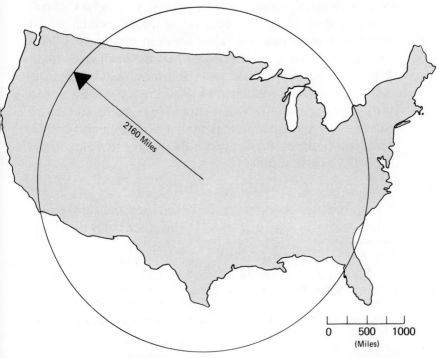

2160 Miles

0		500	1000

(Miles)

Figure VI-3. The size of the moon compared with the continental United States. The moon has ¼ the earth's diameter, and therefore 1/16 the area.

material which on a shorter time-scale would behave as a solid is bound to get crushed under its own weight to settle down to a form of minimum potential energy—a sphere. This is why not only gaseous stars like the Sun, but also solid bodies like the major or terrestrial planets, are spherical.[16]

If the moon were reduced to the previous figure of 748 km in diameter (p. 131), it would not have sufficient gravitational attraction to maintain its spherical shape during meteoritic bombardment and the constant distortion from earth's gravitational pull.[17] In comparison, the two small moons of Mars

---

[16]Z. Kopal, *The Solar System* (New York: Pergamon Press, Inc., 1973), p. 75.

[17]In footnote 16, Kopal states that the central pressure P_c inside a self-gravitating homogeneous configuration of mass with density $\int$ and diameter d is

$$P_c = \frac{\pi}{6} G \int^2 d^2,$$

where G is the gravitational constant (see footnote 18). By inserting the present

are non-spherical, oblong objects because of this lack of sufficient binding mass, and because of heavy bombardment from the asteroid belt which lies between the orbits of Mars and Jupiter. Some scientists believe that the small satellites of Mars will eventually break apart and form rings around the planet. It has long been known that the lunar mass is too small to bind any significant atmosphere to its surface or maintain any surface liquid, as the Apollo explorers verified. Nevertheless, the moon is sufficiently massive to maintain the shape of a beautiful sphere in the sky.

OCEAN TIDES

The attractive gravitational force between the earth-moon system measures 2×10^{16} tons.[18] This force binds the moon

---

lunar values

$$d = 3.479 \times 10^6 \text{m}$$
$$\rho = 3.340 \times 10^3 \text{kg/m}^3$$

into the expression, the actual central pressure is found to be
$$P_c = 4.71 \times 10^9 \text{newtons/m}^2.$$

If the lunar diameter were decreased to 21.5% of its actual value (p. 131), to 7.48×10^5m, the new central pressure P'_c would be
$$P'_c = (.215)^2 P_c$$
$$= 0.046 P_c$$
$$= 2.18 \times 10^8 \text{ newtons/m}^2$$

This is less than the crushing pressure for granite and other similar rocks. Actually the 748 kilometer figure is close to the critical size for sphericity. Others have *estimated* that if a satellite is *at least* 500 kilometers in diameter, its self-gravitation dominates all other cohesive forces, and determines its shape and strength. See E. Smith and K. Jacobs, *Introductory Astronomy and Astrophysics* (Philadelphia: W. B. Saunders Company, 1973), p. 77.

[18] According to Newton's "Universal Law of Gravitation," the attractive force between any two objects of masses m_1 and m_2 separated by a distance d, is given by
$$F = \frac{G m_1 m_2}{d^2}$$
where G is the gravitational constant, equal to
$$G = 6.67 \times 10^{-11} \frac{\text{newton} \times \text{meter}^2}{\text{kilogram}^2}$$
For an earth mass m_1 of 5.98×10^{24} kilograms, a lunar mass m_2 of 7.35×10^{22} kilograms $(0.0123 m_1)$, and a separation of 3.844×10^8 meters, the attractive force is
$$F = 1.98 \times 10^{20} \text{newtons}$$
$$= 4.46 \times 10^{19} \text{pounds}$$
$$= 2.23 \times 10^{16} \text{tons}$$

Suppose the natural gravitational force between the earth and moon were replaced by a steel cable with a tensile strength of 50,000 pounds per square inch. The cable would need to be 531 miles in diameter to provide an equivalent binding force without breaking!

to the vicinity of earth and also is responsible for the ocean tides. Tides actually depend on the change of gravitational force between opposite sides of the earth. Two tidal bulges continually occur on the earth due to the pull of the moon (Fig. VI-3). The side of the earth nearest the moon feels the greatest gravitational pull and bows outward in a high tide. The far side of earth feels the least attraction so that the water flows away from the moon to produce another high tide there. As the moon orbits the spinning earth, two high tides (and also two low tides) occur during approximately each 25-hour period at a given location on the earth. Because of the proximity and size of the moon, the lunar tidal effect is more than twice as great as the sun's tidal effect on the earth. Actual tidal phenomena in any local region depend on shoreline and ocean basin configurations.

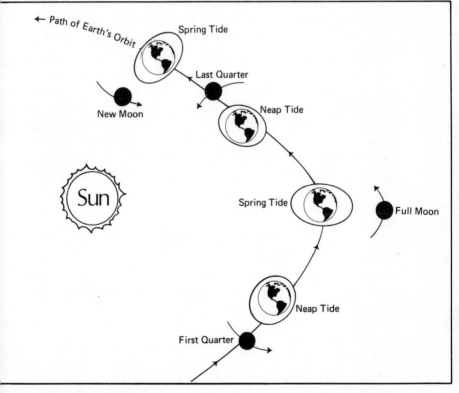

Figure VI-4. Earth tides caused by the moon, appearing on both the near and far side of earth.

The lunar tides have the greatly beneficial effects of cleansing shorelines and diluting stream discharges from land by the large-scale mixing process of currents. These tidal currents regularly scour out shipping channels and keep them open. The high tide permits navigation of waters which are too shallow at other times. In addition, it is now recognized that tides can be tapped as a "lunar energy source" for a world which is rapidly depleting its fossil fuel "inheritance." The rate of tidal-energy dissipation in the earth has been estimated at about 2.6×10^{12} watts, which is close to the world's power consumption.[19]

Historically, the sewer system of the New Testament city, Caesarea of Palestine, was flushed out twice daily by high tides![20] Water pumps in London and mills in New England operated from tidal motion during the eighteenth century.[21] Today, the resource is utilized by a 540-megawatt tidal generating plant on the Rance River in France which harnesses the power of the 28-foot English Channel tides. Thus the large size of the moon helps to provide for man's welfare in his energy needs.

The tidal force felt by the earth depends on the mass of the moon, again proportional to its diameter cubed. A smaller moon would not provide the many useful tidal functions. If, on the other hand, the lunar size were increased to just twice the present diameter, the magnitude of earth tides would be increased by a factor of eight. The resulting tidal waves and earthquakes would show again the futility of trying to improve on the created moon.

FURTHER LUNAR PROPERTIES

This analysis has concerned one unusual property of the moon; namely, its large mass. Numerous other significant lunar features exist for our benefit. The following paragraphs describe some of them.

[19]P. Goldreich, "Tides and the Earth-Moon System," *Scientific American* 226:4 (April, 1972), 43.

[20]Josephus, *The Antiquities of the Jews* 15:9:6.

[21]J. Priest, *Energy for a Technological Society* (New York: Addison-Wesley Publishing Co., 1975), p. 287.

Orbital Angular Momentum

The orbital angular momentum of the moon is defined as the product of its mass, distance from earth, and orbital speed. Again a special lunar origin is indicated by the moon's orbital angular momentum about the earth which exceeds the earth's rotational angular momentum about its own axis. For all other planet-moon systems, the orbital angular momentum of the satellite is a small fraction of the rotational momentum of the planet (Table VI-3).[22] This fact raises severe difficulties with the fission theory of the moon's origin, as discussed on pages 37-42.

Lunar Phases

The systematic phases of the moon are of unending beauty and interest. With a total period of 29½ days, the major phases—new moon, first quarter, full moon, third quarter—are each approximately one week apart. This monthly regularity provides a basis for the calendar which God established at the beginning of time. Notice that the monthly periodicity of the moon's phases is very suitable for time-keeping. If a lunar cycle were as long as a quarter year, it would merely repeat the information provided by the seasons. If, on the other hand, there were as many as 50 lunar phase cycles in a year, keeping track of them would become complicated and cumbersome (see Appendix I).

Lunar Inclination

The plane of the lunar orbit is inclined to the ecliptic (the plane of the earth's orbit about the sun) by 5°9'. The earth's equator is inclined to the ecliptic by 23.5°. Most other satellites are found to be closely located in the planes of their respective planet's equators. If this were true of the earth's moon, little reflected light would ever be provided for the far-Northern and Southern latitudes. The tilt of the lunar orbit to the earth's equator also provides the "harvest moon." This is the next full moon following the autumnal

---

[22]B. G. Marsden, and A. G. W. Cameron, Editors, *The Earth-Moon System.* Proceedings of an International Conference, January 20-21, 1974. Goddard Space Flight Center, NASA (New York: Plenum Press, 1974), p. 165.

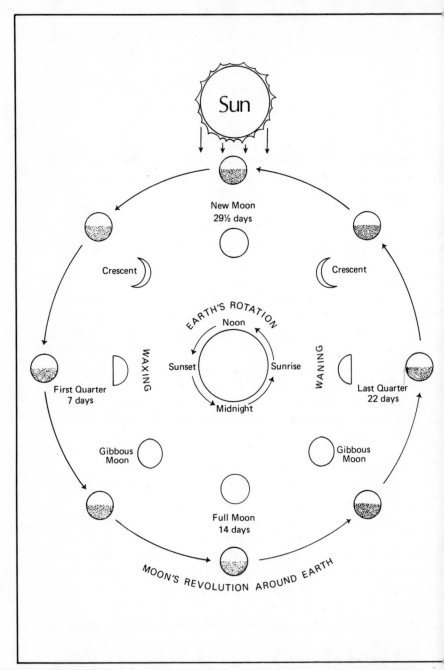

Figure VI-5. The phases of the moon and the corresponding positions of the earth, moon, and sun. The inner figures show the moon as seen from the earth during its phases.

equinox (September 21), when there is the shortest delay in moonrise for several evenings in the Northern hemisphere. On an annual average, the moon rises about 50 minutes later each night. However in September the "harvest moon" rises as little as 10 minutes later on successive nights. Farmers have traditionally enjoyed extra illumination at night to work on their harvest. A compensating delay of the rising of the moon occurs in March, six months after the autumnal equinox.

EARTH — MOON COMPARISONS

	Earth	Moon
eparation	384,404 km = 239,000 miles, or	1.28 light-seconds
iameter	12,742 km = 7,918 miles	3,476 km (0.273 earth)
ass	5.98×10^{24} kg	7.35×10^{22} kg (0.0123 earth)
ensity	5.52 gms/cm³	3.34 gms/cm³ (0.6 earth)
lagnetic field	Yes	No- Remanent magnetism only
tmosphere	Yes	No-No sound, wind, rain, water erosion, twilight
emperature	15.5°C (60°F) Average	204°C (400°F) - lunar day -205°C (-338°F) - lunar night
ater	Covers 75% of the land surface	Trace amounts only
on	Plentiful in oxide forms: Fe_2O_3, Fe_3O_4	Less abundant than earth. Found in pure metallic form
efractory elements (High melting temp) Ca, Ti, Al, U, Zr, Cr		Enriched, compared to earth
olatile elements (Easily melted) o, Na, K		Depleted, compared to earth
urface rock types	Sedimentary 66% Igneous 25% Metamorphic 9%	Igneous 100%
ercent of incident light reflected (albedo)	39%	7%
evolving, orbital period	365.26 days	29.531 days (synodic period) 27.322 days (sidereal period)
otation period	23ʰ 56ᵐ 04ˢ	29.531 days
ravitational acceleration at surface	9.8 m/sec²	1.63 m/sec² (0.17 earth)
rbital eccentricity	0.01673	0.055 (average)
clination of orbit to the ecliptic	0.0 (definition)	4°57'-5°20'
clination of equator to orbit plane	23°27'	6°41'
clination of moon orbit to earth equator		18.5°-28.5°

Table VI-1. Physical Properties of the Earth and Moon (km=kilometers, kg=kilograms, gms=grams, m=meters, sec=seconds, h-m-s=hours, minutes, seconds)

Planet	Satellite	Ratio of satellite mass to planet mass
Earth	Moon	0.0123
Jupiter	Io	4.7×10^{-5}
	Europa	2.56×10^{-5}
	Ganymede	7.84×10^{-5}
	Callisto	5.6×10^{-5}
Saturn	Mimas	6.7×10^{-8}
	Enceladus	1.5×10^{-7}
	Tethys	1.1×10^{-6}
	Dione	1.8×10^{-6}
	Rhea	4×10^{-6}
	Titan	2.41×10^{-4}
	Hyperion	2×10^{-7}
	Iapetus	2.7×10^{-6}
Neptune	Triton	0.002

Table VI-2. A Comparison of Natural Satellite Masses, Relative to Their Planets. Satellites not listed, such as those of Mars, have mass ratios less than 10^{-8}. Note that the earth's moon has the largest mass ratio, as well as being nearest the sun of all the solar system's moons.

Planets with discovered satellites	Ratio of satellite angular momentum to that of the planet-satellite system
Earth	0.829
Mars	Less than 0.001
Jupiter	0.00976
Saturn	0.012
Uranus	0.00612
Neptune	0.113

Table VI-3. An Angular Momentum Comparison for Planets with Satellites. The right column is the ratio of the angular momentum in the orbital motion of the satellite to the total angular momentum of the planet-satellite system.

Orbital Synchronism

The rotation of the moon about its axis and its orbital revolution about the earth have identical periods. The result is that the same side of the moon always faces the earth. This is due to the tidal forces exerted by the earth on the moon. Laser altimeters employed during the Apollo 15 and 16 missions showed that the lunar center of mass is offset from the geometric center by about 2 km toward the earth. Hence, the near side of the moon experiences a larger gravitational force than the far side. The near side is gravitationally "locked in" at its near position. Thus, from a given location on the moon, the earth permanently maintains the same position in the sky, never moving by more than a few degrees due to lunar oscillation, or libration. Photographs of the moon's hidden backside show a hemisphere rich in highly reflective rough terrain. It lacks the large mare lava regions that appear dark on the visible side. If the moon rotated either more rapidly, or more slowly than it actually does, so that the earth was exposed to the far side, the moon would be considerably brighter at this time due to increased reflection. The equal periods of 29½ days for the moon's rotation and revolution thus maintain a dependable brightness, apart from phase changes. The actual lunar albedo (fraction of sunlight reflected from the surface) is 7%, very low when compared with light reflected from the earth (39%), or Venus (76%). If the moon reflected sunlight from a cloudy atmosphere, like Venus, instead of only from its dull dust, a moonlit night would be much brighter, thus interfering with the daily light-darkness cycle.

Conclusion

Some people choose to ignore the significance of these beneficial distinctives of the moon. This approach ultimately leads to the hollow arguments of Woolsey Teller:

> Of what use, then, is this cold, dismal world [moon], with its volcanic craters fifty miles wide and its bleak mountain ranges, moving perpetually around us? And why should there be phases of the moon, varying by degrees from a thin, crescent-shaped strip of light reflected on its surface to an occasional "full" moon thirteen times a year? If the moon, as God-believers insist, was

"intended" to illuminate the night, it is a distinct failure, since the greater part of its time is consumed in reflecting only narrow strips of feeble light and sometimes no light at all![23]

Teller has entirely missed the values of the lunar properties that he superficially mentions. It is truly a pitiful condition to live in a world which is filled with purposeful design, while being blind to God's handiwork.

The multiple evidences of the value and beauty of the moon are not meant in themselves to prove the truth of Christianity. The witness of nature was never intended by God to be a substitute for special revelation. Its function is to remind men of what they already know about God and to activate their consciences with respect to their spiritual responsibility (cf. Rom. 1:18-23). It is hoped that this presentation on lunar distinctives will provide a basis for further study of creation and the perfect Creator, to whom the moon owes its perfect design and whose glory it eloquently declares.

---

[23]W. Teller, *The Atheism of Astronomy*, p. 93.

APPENDIX I

The Dependability and Destiny of the Moon

Following the Flood, God assured the human race that "while the earth remaineth, seedtime and harvest, and cold and heat, and summer and winter, and day and night shall not cease" (Gen. 8:22). This pronouncement of "limited uniformitarianism"[1] guaranteed the dependability of the relative motions of the earth and the sun "while the earth remains." Later promises of God include the moon as well, as will be demonstrated below.

Regardless of how men may react to this great promise, God apparently takes it seriously. Thousands of years after the Flood, at the end of the seventh century B.C., God assured the Jews of Judah and Jerusalem that in spite of the forthcoming and richly deserved Babylonian captivity, the nation of Israel as such could never be destroyed (Jer. 31:35-36; cf. 33:20,25). This was not because of their spiritual and moral qualities as a people (cf. Deut. 7:7-8), but because of God's unconditional and eternal covenant with Abraham (cf. Gen. 12:1-3; Gal. 3:16-18).

Now it is highly significant that the analogy God appealed to for His faithfulness in keeping covenant promises is *the dependability of the sun, moon, and stars.* "Thus saith the Lord, which giveth the sun for a light by day, and the ordi-

---

[1] For an analysis of the Biblical concept of "limited uniformitarianism," see John C. Whitcomb, *The World That Perished* (Grand Rapids: Baker Book House, 1973), pp. 102-04. Cf. Whitcomb and Morris, *The Genesis Flood*, xii, note 1.

nances of the moon and of the stars for a light by night . . . The Lord of hosts is his name: if those ordinances depart from before me, saith the Lord, then the seed of Israel also shall cease from being a nation before me for ever" (Jer. 31:35-36).

Solomon, probably the wisest naturalist between Adam and Christ (cf. 1 Kings 4:29-34), referred to the sun and moon as models of permanence and dependability when he spoke of worship in the future Kingdom Age: "They shall fear thee as long as the sun and moon endure, throughout all generations" (Ps. 72:5). Likewise, Ethan the Ezrahite, a contemporary of Solomon (cf. 1 Kings 4:31), wrote that the Davidic throne would be "established for ever as the moon, and as a faithful witness in heaven" (Ps. 89:37).

In the western world today the moon has lost the significant part it once played as a celestial calendar. The month now serves merely as a convenient time period that is shorter than a season, and it does not even correspond to the approximately 29½ days that elapse between new moons.

In the ancient Near East, however, the moon was very important as a basis for the religious calendar. Thus, in the Old Testament the Hebrew words translated "month"[2] appear over 250 times, frequently in connection with the religious festival of "the new moon."[3] In the name of the Lord, Moses commanded Israel: "in the beginnings of your months, ye shall blow with the trumpets over your burnt-

---

[2]The basic word for "month" in the Hebrew Bible is *ḥōdes̆* (literally, "new," in reference to the new moon). Another word, appearing 13 times, is *yerah̲*, similar to *yāreah̲*—"moon." In the New Testament, the basic Greek word for "month" is *mēn*, appearing 18 times (corresponding to *ḥōdes̆*). The word for "new moon" (*noumēnia*) occurs only in Colossians 2:16.

[3]For a helpful analysis of the use of the moon to establish a calendar in Biblical times, see James L. Boyer, *Chronology of the Crucifixion and the Last Week* (Winona Lake, Ind.: BMH Books, Box 544, 1976). Among the sources used by Boyer are E. J. Bickerman, *Chronology of the Ancient World* (Ithaca, N.Y.: Cornell University Press, 1968); S. Langdon and J. K. Fotheringham, *The Venus Tablets of Ammizaduza* (London: Oxford University Press, 1928), pp. 94-103; R. A. Parker and W. A. Dubberstein, *Babylonian Chronology 626 B.C.–A.D. 95* (Providence, R.I.: Brown University Press, 1956); R. A. Parker, "Ancient Jewish Calendation: A Criticism," *Journal of Biblical Literature,* 63 (1944), 173-76; and J. K. Fotheringham, "Evidence of Astronomy and Technical Chronology for the Date of the Crucifixion," *Journal of Theological Studies,* 35 (1934), 146-62.

offerings, and over the sacrifices of your peace-offerings; that they may be to you for a memorial before your God: I am the Lord your God" (Num. 10:10; cf. 28:11).

The dependability of the moon is not contradicted, but in a sense confirmed by the Biblical references to its future fluctuations of light intensity. The very fact that the moon will not shine during the Day of the Lord, with its global judgments, is set forth as an incomparably great wonder of earth history, which gains in sign value in exact proportion to its normal dependability. This is characteristic of all Biblical sign-miracles. They do not destroy, but rather presuppose a basic uniformity of natural and historical processes.

Eight centuries before Christ, the prophet Joel announced concerning the divine judgments that shall immediately precede the Second Coming of Christ: "the sun and the moon shall be dark, and the stars shall withdraw their shining . . . . for the day of the Lord is great and very terrible; and who can abide it?" (Joel 2:10-11, cf. 3:15). God Himself warns the world: "I will shew wonders in the heavens. . . . the sun shall be turned into darkness, and the moon into blood, before the great and the terrible day of the Lord come" (Joel 2:30-31; quoted by Peter in Acts 2:19-20). Ezekiel applied this coming global judgment to Egypt in particular. "I will cover the sun with a cloud, and the moon shall not give her light" (Ezek. 32:7).

Building upon these warnings of celestial judgment signs at the end of this age, Christ predicted that "there shall be signs in the sun, and in the moon, and in the stars; and upon the earth distress of nations, with perplexity. . . ." (Luke 21:25). More specifically: "Immediately after the tribulation of those days shall the sun be darkened, and the moon shall not give her light . . . and the powers of the heavens shall be shaken" (Matt. 24:29; cf. Mark 13:24-25).

Sixty years later, the Apostle John, exiled to a lonely island in the Aegean Sea, was granted a vision of these unique end-time events: "And I beheld when he had opened the sixth seal, and, lo, . . . the sun became black as sackcloth of hair, and the moon became as blood. . . . and the third part of the sun was smitten, and the third part of the moon . . . so as the third part of them was darkened, and the day shone not for a third part of it, and the night likewise" (Rev. 6:12, 8:12).

Following the Great Tribulation with its unique and astounding suspension of solar and lunar functions in reference to earth-dwellers (cf. Joshua 10:12-13, Hab. 3:11, Luke 23:44-45), the great Kingdom Age will dawn at last.[4] So marvelous will be the glory of the King of Kings at His first appearance upon the earth since His ascension 1,900 years ago that "the moon [Hebrew: $l^e\underline{b}\bar{a}n\hat{a}$, "white one"] shall be confounded, and the sun ashamed, when the Lord of hosts shall reign in mount Zion, and in Jerusalem" (Isa. 24:23). From another perspective, however, the Kingdom Age will be such a contrast to the dreadful darkness that just precedes it that "the light of the moon [$l^e\underline{b}\bar{a}n\hat{a}$] shall be as the light of the sun, and the light of the sun shall be sevenfold, as the light of seven days, in the day that the Lord bindeth up the breach of his people, and healeth the stroke of their wound" (Isa. 30:26).

At the end of the Kingdom Age, John describes the moon's destiny: "And I saw a great white throne, and him that sat on it, from whose face the earth and the heaven fled away; and there was found no place for them. . . . And I saw a new heaven and a new earth: for the first heaven and the first earth were passed away; and there was no more sea. And I John saw the holy city, new Jerusalem, coming down from God out of heaven, ... And the city had no need of the sun, neither of the moon, to shine in it: for the glory of God did lighten it, and the Lamb is the light thereof. . . . for there shall be no night there" (Rev. 20:11, 21:1-2, 23, 25; cf. Isa. 60:19-20).

Thus the moon, like other heavenly bodies, will continue to fulfill its three God-ordained purposes of *illuminator, time regulator,* and *sign* (Gen. 1:14) as long as the earth remains (cf. Gen. 8:22), even into the Kingdom Age, with an eschatological emphasis upon its sign function. But the moon is not essential to human existence. And it is certainly not a god to be worshiped! Beautiful, complex, and functionally impor-

---

[4]For a definitive study of this coming age, see Alva J. McClain, *The Greatness of the Kingdom* (Winona Lake, Ind.: BMH Books, 1959).

tant though it may be to the inhabitants of the earth, it is infinitely inferior to the God who created it. "To whom then will ye liken me, or shall I be equal? saith the Holy One. Lift up your eyes on high, and behold who hath created these things, that bringeth out their host by number: he calleth them all by names by the greatness of his might, for that he is strong in power; not one faileth. . . . Look unto me, and be ye saved, all the ends of the earth: for I am God, and there is none else" (Isa. 40:25-26, 45:22).

APPENDIX II

Moon Worship --
A Spiritual Disaster

Currently popular schemes for accommodating Genesis 1 to cosmic evolutionism have produced a double loss for the Christian world. First, there has been a loss of loyalty and sensitivity to the God-honored and time-honored principles of interpreting the Bible (Biblical hermeneutics). Second, there has been a resulting loss of essential insights concerning the glory and sovereignty of God, not only in the opening statements of Scripture but also throughout the Bible, with staggering theological, philosophical, ethical, literary, and artistic implications for modern culture and society.[1]

It is the purpose of this supplementary study to explore some of the major *theological* implications of the Genesis account of the creation of the sun, moon and stars. What message did God intend to communicate to the mind of man, both ancient and modern, through this record of astronomical origins? Have we suffered a major loss by failing to hear that message? What desperately dangerous attitudes was that message intended to replace?

THE THEOLOGICAL SIGNIFICANCE OF THE MOON'S CREATION ON THE FOURTH DAY

It has not been generally understood that *the order of*

[1]See Francis A. Schaeffer, *The God Who is There* (Chicago: InterVarsity Press, 1968), and Os Guinness, *The Dust of Death* (Downers Grove, Ill.: InterVarsity Press, 1973).

events in Genesis 1 is deliberate and meaningful, and that all alternative views to the traditional view of a comparatively recent creation week share a common denial of this divine order.[2] While claiming to allow the opening chapter of the Bible to speak for itself, proponents of these alternative views are in reality reversing the created order of the earth and the solar system to accommodate the uniformitarian consensus among contemporary astronomers. One result of this reversal is the tragic eclipse of a profound message that permeates the entire Old Testament: *the Creator is infinitely superior to the creation, including the astronomical creation. Therefore no visible heavenly body, including the moon, may be worshiped.*

This profound theological principle is confirmed and demonstrated Biblically by the fact that the moon was directly and instantaneously brought into existence apart from pre-existent materials by a mere spoken word of the absolutely unique, omnipotent, omniscient, holy and transcendent God of the universe, Jehovah of Israel, who alone is to be worshiped by men. Furthermore—and this neglected fact is theologically crucial—the infinite inferiority of the moon to the true God of creation is fixed by its having been created *after* the creation of the earth and its vegetation. In the words of a widely read professor of the history of science:

> In the first chapter of Genesis it is made evident that absolutely nothing, except God, has any claim to divinity; even the sun and moon, supreme gods of the neighboring peoples, are set in their places between the herbs and the animals and are brought into the service of mankind.[3]

Idolatry is far more subtle and complex—and to that extent less "honest" and open—in our generation than in the

---

[2] For critiques of the Gap Theory of Genesis 1:2, the Day-Age Theory, the Revelatory Theory and the Framework Hypothesis, see references in Chapter III, note 28. For exegetical evidence that the initial creation of the sun, moon and stars occurred on the fourth day of creation week, and thus only two days before the creation of mankind, see Chapter III, note 31. (This fact alone destroys all the alternatives to the recent-creation position, for these compromise positions were postulated specifically to accommodate the supposed vast antiquity of the astronomical bodies.) For the absolute date of creation week, see Chapter III, note 48.

[3] R. Hooykaas, *Religion and the Rise of Modern Science,* p. 8.

ancient world. Therefore a study of the worship of the sun and moon in the "cradle of civilization" and a survey of the Old Testament denunciation of that worship may seem to be utterly irrelevant and remote to the modern reader. It is clear to discerning students, however, that while the forms and objects of pagan worship may change through the centuries, the underlying rejection of the witness of the Spirit of God to the human heart through general and special revelation remains the same. Modern minds, no less than ancient minds, sense that the Creator who is revealed in the opening chapter of the Bible is vastly different in His attributes and in His demands upon men than any other god ever imagined. This is why our understanding of the manner and the relative chronology of the moon's creation can never be a peripheral issue, even to "modern" men.

THE OLD TESTAMENT AND MOON WORSHIP

Assuming a date for the Book of Job around 2100 B.C.,[4] we have in this remarkable document the earliest inspired record of man's thinking about the moon. Bildad the Shuhite asked Job: "If even the moon has no brightness and the stars are not pure in His sight, how much less man. . . .!" (Job 25:5-6 NASB; cf. 15:15). Job himself confirmed this concept of the infinite superiority of God to the moon, and insisted that worship of the sun or moon would rightly incur judgment because of the blasphemy involved in such idolatry. "If I beheld the sun when it shined, or the moon walking in brightness; and my heart hath been secretly enticed, or my mouth hath kissed my hand [in the homage of worship]: this also were an iniquity to be punished by the judge: for I should have denied the God that is above" (Job 31:26-28).

Several centuries later (c. 1405 B.C.), when Moses was preparing Israel for entrance into the land of Canaan, stern warnings were issued concerning the desperate danger of being influenced by the solar, lunar, and astral worship cults of the nations which had lived there for centuries, "lest thou lift

---

[4]In all of the extensive theological debates recorded in the Book of Job, no mention is made of God's epochal covenant with Abraham. For additional evidence in support of this early date for Job, see J. C. Whitcomb, *From the Creation to Abraham* (Winona Lake, Ind.: BMH Books, 1977), note 8.

up thine eyes unto heaven, and when thou seest the sun, and
the moon, and the stars, even all the host of heaven, should-
est be driven to worship them, and serve them, which the
Lord thy God hath divided unto all nations under the whole
heaven" (Deut. 4:19; cf. 17:3 and Amos 5:36).

After the establishment of the theocratic kingdom under
David, God made a promise to His people that has caused
some perplexity: "Behold, he that keepeth Israel shall neither
slumber nor sleep. The Lord is thy keeper: the Lord is thy
shade upon thy right hand. The sun shall not smite thee by
day, *nor the moon by night.* The Lord shall preserve thee
from *all evil. . . ."* (Ps. 121:4-7). While it is true that sun-
stroke was at times a special danger in the Near East (cf. II
Kings 4:18-20; Jonah 4:8; Isa. 49:10), the heat of the sun
could hardly qualify as one of Israel's major threats, to say
nothing of "the moon by night."[5] It seems more appropriate,
therefore, to consider this promise within the context of sun
and moon worship. Idolatrous people were presumably *afraid*
that a neglect of proper sacrifice to these heavenly deities
would result in their being somehow smitten during the day
by the sun or during the night by the moon. Thus, Jeremiah
could also reassure Israel: "Thus saith the Lord, Learn not
the way of the heathen, and *be not dismayed at the signs of
heaven; for the heathen are dismayed at them"* (Jer. 10:2).

About 700 B.C., seven centuries after Israel had entered
the Promised Land, "crescent ornaments" (Isa. 3:18), shaped
like the moon and directly inviting worship of the moon (cf.
Judges 8:21,26), were popular with "the daughters of Zion"
whose lifestyles were influenced by Near Eastern idolatry
more than by their Lord.

Soon after the days of King Hezekiah and Isaiah the
prophet, during the seventh century B.C., celestial idolatry

---

[5] Derek Kidner admits that little is understood concerning "the effects of the
moon on certain people." But he still feels that "some kinds of mental disturbance vary with its phases. Not all popular belief on the subject is unfounded"
(*Psalms 73-150*, London: InterVarsity Press, 1975, p. 432). H. C. Leupold con-
cedes only that "the expression does allow for the fact that such a notion was
spread abroad in Israel" (*Exposition of Psalms*; Grand Rapids: Baker Book House,
1969, p. 870). Certainly the fact that the term "lunatick" (*selēniazomai* = "moon
struck") appears in Matthew 4:24 and 17:15 proves only that this popular notion
was current in New Testament times as well.

swept into Judah like a flood. King Manasseh (690-640 B.C.) actually "worshipped all the host of heaven, and served them. And he built altars . . . . for all the host of heaven in the two courts of the house of the Lord" (II Kings 21:3-5). When Josiah inherited the throne, he attempted to purge the land of these influences and especially "the idolatrous priests . . . who burned incense unto Baal, to the sun, and to the moon . . . ." (II Kings 23:5).

But Josiah's reforms were too superficial and too late. Radical forms of idolatry had deeply infected the hearts of the vast majority of Jews. Therefore Josiah's great contemporary, the "weeping prophet" Jeremiah (cf. Jer. 9:1, 13:17), was commanded by God to cease praying for the nation (Jer. 7:16). Judah was almost totally corrupted by submitting to the all-encompassing influence of Near Eastern idolatry (including the worship of the moon), and would therefore be destroyed and deported by the cruel armies of Nebuchadnezzar, the king of Babylon. "At that time, saith the Lord, they shall bring out the bones of the kings of Judah . . . and the bones of the inhabitants of Jerusalem, out of their graves: And they shall spread them before the sun, and the moon, and all the host of heaven, whom they have loved, . . . served, . . . after whom they have walked, and . . . sought, and whom they have worshipped: they shall not be gathered, nor be buried; they shall be for dung upon the face of the earth" (Jer. 8:1-2).

What the warnings of Moses, Isaiah and Jeremiah did not accomplish for the people of Israel, the Babylonian Captivity *did* accomplish, at least superficially. The Jews became known henceforth as "the people of the Book," and gross idolatry such as moon worship receded into a mere memory of their pre-exilic past, enshrined forever in the inspired records of the Old Testament.

WORSHIP OF THE MOON IN THE ANCIENT NEAR EAST

What exactly were these enormous influences toward the worship of the moon that so threatened Job and his contemporaries in the late third millennium B.C. and the people of Israel for over 800 years after their exodus from Egypt?

The science of archeology has demonstrated the deification of the moon from early Sumerian times (third millennium B.C.) to Islamic times throughout western Asia, the cradle of postdiluvian civilization. The great Sumerian city of Ur, in lower Mesopotamia, was especially devoted to the worship of the moon, under the name of *Nanna* or *Nannar,* long before the time of Abraham. Surprisingly, *Nanna* was thought to be the father of the sun-god *Shamash* (though he himself was the son of a yet greater god, *Enlil*).[6]

Soon after 2000 B.C., the moon temple at Ur was greatly enlarged by Ur-Nammu, first king of the great Third Dynasty of Ur. Fourteen hundred years later King Nabonidus of Babylon restored this temple. A cuneiform inscription has been found there in which "Nabonidus concluded with a dedication to Nannar, lord of the gods of heaven and earth, and a prayer for the life of himself and of his son Belshazzar."[7] This temple (or ziggurat) remains today as the best preserved of the many that can still be seen in Mesopotamia (see Fig. 1).[8]

The Akkadians called the moon-god by the name *Sin,* "the lamp of heaven and earth,"[9] "the king of all gods,"[10] and "the Divine Crescent."[11] The crescent is the familiar symbol

---

[6]Cf. "The Code of Hammurabi," trans. by Theophile J. Meek, *Ancient Near Eastern Texts Relating to the Old Testament* [hereinafter referred to as *A.N.E.T.*], ed. by James B. Pritchard (Princeton: Princeton University Press, 3rd ed., 1969), p. 164, note 10; G. R. Driver and John C. Miles, *The Babylonian Laws* (London: Oxford University Press, 1955), II, p. 122 (lines 14-15), p. 298 (lines 41-42); and general analysis in Jack Finegan, *Light From the Ancient Past,* 2nd ed. (Princeton: Princeton University Press, 1959), p. 566, note 1.

[7]Jack Finegan, *Light from the Ancient Past,* p. 50.

[8]Photographs of this ziggurat may be seen in Finegan, *Light From the Ancient Past,* Figure No. 19 (following page 26); Martin A. Beek, *Atlas of Mesopotamia* (New York: Thomas Nelson and Sons, 1962), Figure No. 273, p. 141; and Charles F. Pfeiffer and Howard F. Vos, *The Wycliffe Historical Geography of Bible Lands* (Chicago: Moody Press, 1967), p. 18.

[9]"Psalm to Marduk," trans. by Ferris J. Stephens, *A.N.E.T.,* p. 390. Cf. also the "Hymn to the Moon-God," and the "Prayer to the Moon-God," also translated by Stephens, pp. 385-86.

[10]Called thus by the mother of King Nabonidus. Cf. *A.N.E.T.,* p. 560.

[11]Thus named by Nabonidus himself. Cf. *A.N.E.T.,* pp. 562-63.

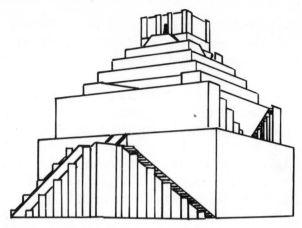

Figure A II-1. A reconstruction of a Babylonian ziggurat from information discovered through excavations. "A text from Nippur addresses the
temple tower in these words: 'Great mountain of Enlil, whose peak
reaches the sky . . . .' These texts might also be quoted as so many parallels to Genesis 11:4 and in support of the theory that the Biblical writer
correctly states the [religious] concept of the ziggurat . . . ." Martin A.
Beek, <u>Atlas of Mesopotamia</u> (New York: Nelson, 1962), p. 143.

for *Sin* in Mesopotamian art.[12] When the far-off Hittites appealed to the Sun-god in their hours of special need, they
referred to him as "the favorite son of Sin."[13] The great
Hammurabi (c. 1700 B.C.) introduced himself in his famous
Code as "the descendant of royalty, whom Sin begat."[14]

It is fascinating to trace the presence of moon worship
along the routes followed by Abraham and his descendants.
Called by the true Creator of the moon to leave his home-
land, Abraham came to the city of Harran in northern Mesopotamia (today on the southern border of southeastern Turkey). To judge from the later prominence of moon worship
in this city, "the Divine Crescent" was probably honored in
his day, too. Like the Apostle Paul who received a similar call

---

[12]For photographic representations, cf. James B. Pritchard, *The Ancient Near
East in Pictures*, 2nd ed. (Princeton: Princeton University Press, 1969), No. 453
(p. 156) and No. 519 (p. 176).

[13]"Prayer of Kantuzilis for Relief from his Sufferings," trans. by Albrecht
Goetze, *A.N.E.T.*, p. 400.

[14]Cf. *A.N.E.T.*, p. 164.

from the Creator of heaven and earth over 2,000 years later, Abraham's spirit must have been "provoked within him as he was beholding the city full of idols" (cf. Acts 17:16, NASB, in reference to Athens).

Leaving Harran about 2090 B.C.,[15] Abraham continued southward toward Canaan, doubtless encountering signs of the Ugaritic moon deity along the way, whose name was *Yarikh.* At the city of Hazor, north of the Sea of Galilee, archeologists have discovered the remains of an ancient Canaanite shrine dedicated to the full and crescent moons (Fig. 2). Proceeding on into Egypt because of a famine in Canaan (to be followed later by Jacob's entire family), he may well have learned of *Khonsu,* the moon-god of Thebes.[16]

As millions of Abraham's descendants reentered the Promised Land under the leadership of Joshua, they were confronted with the fortress-city of Jericho, which as its very name indicates (*yᵉrîhô,* cf. *yārēaḥ,* "moon") was probably dedicated to the Semitic moon-god.[17]

Eight centuries later, as the Jews were deported to Babylon because of their persistent indulgence in moon worship (among other reasons), it is fascinating to ponder that moon worship played a significant role also in the final collapse of the Neo-Babylonian empire. Because his mother Adad-guppi was a devoted priestess of the moon-god *Sin* at Harran,[18] King Nabonidus (who may have married a daughter of Nebuchadnezzar) clashed constantly with the Marduk priests

---

[15]For a chronological analysis of the patriarchal era, cf. J. C. Whitcomb, *Old Testament Patriarchs and Judges* (Winona Lake, Ind.: BMH Books, 1965).

[16]Cf. James B. Pritchard, *The Ancient Near East in Pictures,* No. 563 (p. 188). This inscription was found at Karnak. Cf. also J. Gwyn Griffiths, "Osiris and the Moon in Iconography," *The Journal of Egyptian Archaeology* 62 (1976), 153-59. For the recent discovery of a 20-ton, pre-Columbian circular stone bas-relief of the Aztec moon goddess Coyolxauhqui beneath a street in Mexico City, see *Time* 111:16 (April 17, 1978), 44.

[17]Cf. James Mathisen, "Moon," *Wycliffe Bible Encyclopedia,* ed. by Pfeiffer, Vos and Rea (Chicago: Moody Press, 1975), p. 1148.

[18]"The Mother of Nabonidus," trans. by A. Leo Oppenheim, *A.N.E.T.,* pp. 560-62.

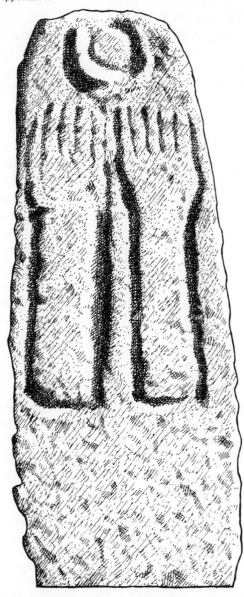

Figure A II-2. A foot-high pillar engraved with hands raised toward the full and crescent moon. Discovered in 1956 in excavations at Hazor, northern Palestine, and dating to about 1400 B.C. (From Yigael Yadin, Hazor: The Rediscovery of a Great Citadel of the Bible [New York: Random House, 1975], pp. 43-47 and cover design.)

Sketch by Arthur Davis

in his capital city of Babylon because of his insistence on introducing moon worship there. Finally, in 553 B.C., he left Babylon, "entrusted the kingship" to his son Belshazzar,[19] and moved to Tema in northwestern Arabia because it was an

---

[19]"Verse Account of Nabonidus," trans. by A. Leo Oppenheim, *A.N.E.T.*, p. 313; cf. "Nabonidus and His God," pp. 562-63.

ancient center of worship of the moon-god.[20]

This royal and priestly clash over the issue of moon worship resulted in Babylon being left in the hands of Belshazzar, an utter profligate (cf. Dan. 5), and thus open to the assault of Cyrus the Great and his Medo-Persian armies in 539 B.C.[21]

CONCLUSION

The prominence of moon worship in the experience of Abraham and his descendants down to the time of Cyrus serves to clarify the message of the first chapter of Genesis which was given to Israel through Moses just before they returned to Canaan. The message was that the moon (as well as all other astronomical bodies) was *not* a god to be worshiped. In fact, so far from being a deity, it was not even the offspring of a god or goddess (as was claimed for *Nanna/Sin).* Instead—and this is utterly and permanently devastating to all forms of moon worship—it was created instantly, *ex nihilo* and subsequent to an earth already carpeted with vegetation by the incomparable God of Israel.

In conclusion, it must be recognized that if the modern theistic-evolution interpretation of the origin of the moon had been adopted by ancient Israel, the message of Genesis would have been reduced to the level of just one more Near Eastern mythological cosmogony with little or no impact upon idolatry-prone ancients. This profoundly important fact must be weighed carefully by theologians who are tempted to mold the opening chapter of the Bible into conformity with currently popular naturalistic theories of origins.

---

[20]Cf. Julius Lewy, "The Late Assyro-Babylonian Cult of the Moon and Its Culmination at the Time of Nabonidus," *Hebrew Union College Annual,* 19 (1945-46), pp. 405-89. Lewy claims that the name Sinai points to the moon deity, and that such Assyrian kings as Sennacherib (Sin-ahhe-eriba) and Sinsharishkun (second son of Ashurbanipal) were dedicated to the moon (p. 455). With regard to Nabonidus, "Since the worshippers of the moon-god were numerous in all his lands, particularly among the Western Semites settled both in Western and Eastern parts of the empire, we may well assume that he could reasonably expect to be supported in his policy [of promoting "The Era of the Moon God"] by a considerable fraction of his subjects" (p. 487).

[21]Cf. J. C. Whitcomb, *Darius the Mede: The Historical Chronology of Daniel* (Nutley, N.J.: Presbyterian and Reformed Publishing Co., 1963), p. 70.

APPENDIX III

The Bible and Science -
A Spectrum of
Representative Writings
Among Christian Theologians
and Scientists

I. REPRESENTATIVES OF THE DOUBLE-REVELATION PERSPECTIVE

(The literality and historicity of Genesis is in some way qualified by uniformitarianism and/or evolutionism. For an explanation of this perspective, see Chapter III, pp. 54-55.)

A. Scientists

Albert, Jerry D., "A Biochemical View of Life," *JASA* 29:2 (June, 1977), 81-84.

Anderson, V. Elving, "Evangelicals and Science," in *The Evangelicals*, ed. by D. F. Wells and J. D. Woodbridge (Nashville: Abingdon Press, 1975), especially pp. 260-66.

Andrews, George W., "Geology," in *Christ and the Modern Mind*, ed. by Robert W. Smith (Downers Grove, Ill.: Inter-Varsity Press, 1972), pp. 267-70.

Aulie, Richard P., "The Doctrine of Special Creation," *JASA* 27:1 (March, 1975) 8-11, and following issues.

Ault, Wayne V., "Flood (Genesis)," *The Zondervan Pictorial Encyclopedia of the Bible* (Grand Rapids: Zondervan Publishing House, 1975), II, p. 563.

Bube, Richard H., *The Encounter Between Christianity and Science* (Grand Rapids: Wm. B. Eerdmans Publishing Co.,

1968), p. 107; *The Human Quest* (Waco, Texas: Word Books, Inc., 1971), pp. 206-09; and "Inerrancy, Revelation and Evolution," *JASA* 24:2 (June, 1972), 81,83.

Bullock, Wilbur L., "Evolution Versus Creation—In Retrospect and Prospect," *Gordon Review* (Summer, 1959), 79.

Cuffey, Roger J., "Paleontological Evidence and Organic Evolution," *JASA* 24:4 (December, 1972), 166,71.

De Vries, John and Donald C. Boardman, *Essentials of Physical Science* (Grand Rapids: Wm. B. Eerdmans Publishing Co., 1958), p. 304.

Dye, David L., *Faith and the Physical World* (Grand Rapids: Wm. B. Eerdmans Publishing Co., 1966), p. 135.

Haas, J. W., Jr., "Biogenesis: Paradigm and Presupposition," *JASA* 27:4 (December, 1975), 152-55, and "What Christian Colleges Teach About Creation," *Christianity Today* 21:18 (June 17, 1977), 8-11.

Hearn, Walter R., and Richard A. Hendry, "The Origin of Life," in *Evolution and Christian Thought Today*, ed. by Russell L. Mixter (Grand Rapids: Wm. B. Eerdmans Publishing Co., 1959), pp. 67-70.

Herrmann, Robert L., "Implication of Molecular Biology for Creation and Evolution," *JASA* 27:4 (December, 1975), 159.

Howkins, Kenneth G., *The Challenge of Religious Studies* (Downers Grove, Ill.: InterVarsity Press, 1973), pp. 111-15.

Jeeves, Malcolm A., *The Scientific Enterprise and Christian Faith* (Downers Grove, Ill.: InterVarsity Press, 1969), p. 103.

Jones, D. Gareth, "Evolution: A Personal Dilemma," *JASA* 29:2 (June, 1977), 73-76.

Kessel, Edward L., "Let's Look at the Facts, Without Bent or Bias," in *The Evidence of God in an Expanding Universe*, ed. by John C. Monsma (New York: G. P. Putnam's Sons, 1958), p. 52.

Kulp, J. Laurence, "The Christian Concept of Uniformity in the Universe," *His Magazine* 12:8 (May, 1952), 15-24.

Lever, Jan, *Creation and Evolution* (Grand Rapids: Kregel, 1958), p. 21.

Mackay, Donald M., *The Clockwork Image* (Downers Grove, Ill.: InterVarsity Press, 1974), pp. 62,69,92.

Mixter, Russell L., "Man in Creation," *Christian Life* 23:6 (October, 1961), 25.

Reid, James, *God, the Atom, and the Universe* (Grand Rapids: Zondervan Publishing House, 1968), pp. 107-09, 173.

Schweitzer, George K., "The Origin of the Universe," in *Evolution and Christian Thought Today*, ed. by Russell L. Mixter (Grand Rapids: Wm. B. Eerdmans Publishing Co., 1959), pp. 34,35,48.

Spanner, D. C., *Creation and Evolution* (Grand Rapids: Zondervan Publishing House, 1968), pp. 9-13, 56-61.

van de Fliert, J. R., "Fundamentalism and the Fundamentals of Geology," *JASA* 21:3 (September, 1969), 69-80.

van der Ziel, *The Natural Sciences and the Christian Message* (Minneapolis: T. S. Denison and Co., 1960), p. 218.

Willis, David L., "Creation and/or Evolution," *JASA* 29:2 (June, 1977), 68-72.

B. Theologians and Philosophers

Byrt, John, "The Roles of the Bible and Science in Understanding Creation," *Faith and Thought* 103:3 (1976), 158-88.

Kline, Meredith G., "Because It Had Not Rained," *Westminster Theological Journal* 20:2 (May, 1958), 146-57.

Kuschke, Arthur W., Jr., Review of *The Genesis Flood* in *The Westminster Theological Journal* 24:2 (May, 1962), 221-23.

Leith, Thomas H., "Galileo and the Church: Tensions with a Message for Today," *JASA* 25:4 (December, 1973), 156.

Ramm, Bernard, *The Christian View of Science and Scripture* (Grand Rapids: Wm. B. Eerdmans Publishing Co., 1954), p. 104.

Richardson, Alan, *The Bible in An Age of Science* (Philadelphia: The Westminster Press, 1961), p. 167.

Ridderbos, N. H., *Is There A Conflict Between Genesis 1 and Natural Science?* (Grand Rapids: Wm. B. Eerdmans Publishing Co., 1957), p. 46.

Stott, John R. W., *Understanding the Bible* (Glendale, Calif.: Gospel Light Pub., 1972), pp. 62-65.

Taylor, Kenneth N., *Evolution and the High School Student* (Wheaton, Ill.: Tyndale House Pub., 1969), pp. 50-54.

II. REPRESENTATIVES OF THE RECENT-CREATION PERSPECTIVE

(The full historicity of the early chapters of Genesis acknowledged)

A. Scientists

Armstrong, Harold L., "Comets and a Young Solar System," in George F. Howe, *Speak to the Earth* (Nutley, N.J.: Presbyterian and Reformed Publishing Co., 1975), pp. 327-30.

Barnes, Thomas G., *Origin and Destiny of the Earth's Magnetic Field* (San Diego: Institute for Creation Research, 1973).

Bliss, Richard B., *Origins: Two Models* (San Diego: Institute for Creation Research, 1976).

Brown, Robert H., "C-14 Age Profiles For Ancient Sediments and Peat Bogs," *Origins* [Geoscience Research Institute, Loma Linda University, Loma Linda, Calif. 92354] 2:1 (1975) 6-18.

Burdick, Clifford L., "Streamlining Stratigraphy," in Walter E. Lammerts, ed., *Scientific Studies in Special Creation* (Nutley, N.J.: Presbyterian and Reformed Publishing Co., 1971), pp. 125-35; and "The Empire Mountains—A Thrust Fault?" in George Howe, ed., *Speak to the Earth* (Nutley, N.J.: Presbyterian and Reformed Publishing Co., 1975), pp. 376-87.

Chittick, Donald E., "Carbon 14 Dating of Fossils" and "Dating the Earth and Fossils," in *A Symposium on Creation II* ed. by Donald W. Patten (Grand Rapids: Baker Book House, 1970), pp. 44-74.

Clark, Harold W., *Fossils, Flood, and Fire* (Escondido, Calif.: Outdoor Pictures, 1968).

Clark, Robert T., and James D. Bales, *Why Scientists Accept Evolution* (Nutley, N.J.: Presbyterian and Reformed Publishing Co., 1966).

Clementson, Sidney P., "A Critical Examination of Radioactive Dating of Rocks," in George F. Howe, ed., *Speak to the Earth* (Nutley, N.J.: Presbyterian and Reformed Publishing Co., 1975), pp. 365-75.

Coffin, Harold G., *Creation—Accident or Design?* (Washington, D.C.: Review and Herald Publishing Assoc., 1969).

Cook, Melvin A., *Prehistory and Earth Models* (London: Max Parrish and Co., 1966).

Culp, G. Richard. *Remember Thy Creator* (Grand Rapids: Baker Book House, 1975).

Davidheiser, Bolton, *Evolution and Christian Faith* (Nutley, N.J.: Presbyterian and Reformed Publishing Co., 1969).

Dillow, Joseph C., "The Catastrophic Deep-Freeze of the Beresovka Mammoth," *CRSQ* 14:1 (June, 1977), 5-13; and "The Canopy and Ancient Longevity," *CRSQ* 15:1 (June, 1978).

DeYoung, Donald B., "Geochemistry of the Stable Isotopes," *CRSQ* 11:1 (June, 1974), 32-36.

Enoch, H., *Evolution or Creation*, rev. ed. (Grand Rapids: Evangelical Press, 1976).

Flori, Jean and Henri Rasoloformasoandro, *Evolution or Creation?* (Editions S.D.T., 77190 Dammarie les Lys, France, 1974).

Frair, Wayne, "Life in a Test Tube," in Walter E. Lammerts, ed., *Why Not Creation?* (Nutley, N.J.: Presbyterian and Reformed Publishing Co., 1970), pp. 268-82.

Gentry, Robert V., "Cosmological Implications of Extinct Radioactivity from Pleochroic Halos," in Walter E. Lammerts, ed., *Why Not Creation?* (Nutley, N.J.: Presbyterian and Reformed Publishing Co., 1970), pp. 106-13; and "Radiohalos in Coalified Wood: New Evidence Relating to the Time of Uranium Introduction and Coalification," *Science* 194:4262 (October 15, 1976), 315-18.

Gish, Duane T., *Evolution: The Fossils Say No!* and *Speculations and Experiments Related to the Origin of Life (A Critique)* (San Diego: Institute for Creation Research [2717 Madison Avenue, San Diego, Calif. 92116], 1972).

Howe, George F., "Paleobotanical Evidences for a Philosophy of Creationism," in Walter E. Lammerts, ed., *Why Not Creation?* (Nutley, N.J.: Presbyterian and Reformed Publishing Co., 1970), pp. 230-42, and three articles in Walter E. Lammerts, ed., *Scientific Studies in Special Creation* (Nutley, N.J.: Presbyterian and Reformed Publishing Co., 1971), pp. 206-28, 243-57, 285-98.

Klotz, John W., *Genes, Genesis and Evolution* (St. Louis: Concordia Publishing House, 1955); "Assumptions in Science and Paleontology," in Paul A. Zimmerman, ed., *Rock Strata and the Bible Record* (St. Louis: Concordia Publishing House, 1970), pp. 24-39.

Kofahl, Robert E., *The Creation Explanation* (Wheaton, Ill.: Harold Shaw Publishers, 1975).

Lammerts, Walter E., "On the Recent Origin of the Pacific Southwest Deserts," in George F. Howe, *Speak to the*

Earth (Nutley, N.J.: Presbyterian and Reformed Publishing Co., 1975), pp. 314-23; and "Discoveries Since 1859 Which Invalidate the Evolution Theory," in Walter E. Lammerts, ed., *Why Not Creation?* (Nutley, N.J.: Presbyterian and Reformed Publishing Co., 1970), pp. 248-67.

Marsh, Frank Lewis, *Life, Man and Time* (Escondido, Calif.: Outdoor Pictures, 1967).

Moore, John N., *Biology: A Search for Order in Complexity*, edited with Harold S. Slusher (Grand Rapids: Zondervan Publishing House, 1973), and *Questions and Answers on Creation/Evolution* (Grand Rapids: Baker Book House, 1976). See Zola Levitt, *Creation: A Scientist's Choice* (Wheaton: Victor Books, 1976), for an analysis of Moore's writings.

Morris, Henry M., *The Genesis Flood*, co-authored with John C. Whitcomb (Nutley, N.J.: Presbyterian and Reformed Publishing Co., 1961), and numerous other works available through the Institute for Creation Research (2716 Madison Avenue, San Diego, Calif. 92116) including *Scientific Creationism* (1974).

Mulfinger, George, "Examining the Cosmogonies—A Historical Review," in Walter E. Lammerts, ed., *Why Not Creation?* (Nutley, N.J.: Presbyterian and Reformed Publishing Co., 1970), pp. 39-66; "Critique of Stellar Evolution," in George F. Howe, ed., *Speak to the Earth* (Nutley, N.J.: Presbyterian and Reformed Publishing Co., 1975), pp. 409-46; and with Emmit L. Williams, *Physical Science for Christian Schools* (Bob Jones University Press, Greenville, S.C. 29614, 1975).

Nevins, Stuart E., "Interpreting Earth History," *JCR* 1:1 (Summer, 1974), 28-34; three articles in George F. Howe, ed., *Speak to the Earth* (Nutley, N.J.: Presbyterian and Reformed Publishing Co., 1975), pp. 16-59; 211-52; "Continental Drift, Plate Tectonics, and the Bible," ICR Impact Series No. 32 (Institute for Creation Research, 2716 Madison Ave., San Diego, Calif. 92116, Feb., 1976), and "The Origin of Coal," ICR Impact Series No. 41 (November, 1976).

Ritland, Richard M., *A Search for Meaning in Nature* (Mountain View, Calif.: Pacific Press Publishing Association, 1970).

Rodabaugh, David J., "The Queen of Science Examines the King of Fools," *CRSQ* 12:1 (June, 1975), 14-18; "Probability and the Missing Transitional Forms," *CRSQ* 13:2 (September, 1976), 116-19.

Rusch, Wilbert H., Sr., "Analysis of So-Called Evidences of Evolution," in Paul A. Zimmerman, ed., *Creation, Evolution, and God's Word* (St. Louis: Concordia Publishing House, 1972), pp. 36-67; "Human Fossils," in Paul A. Zimmerman, ed., *Rock Strata and the Bible Record* (St. Louis: Concordia Publishing House, 1970), pp. 133-77.

Slusher, Harold S., *Critique of Radiometric Dating* (San Diego: Institute for Creation Research, 1973), and "Some Astronomical Evidences for a Youthful Solar System," in George F. Howe, ed., *Speak to the Earth* (Nutley, N.J.: Presbyterian and Reformed Publishing Co., 1975), pp. 401-08.

Tinkle, William J., *Heredity* (Grand Rapids: Zondervan Publishing House, 1970).

Watson, David C. C., *Myths and Miracles* (H. E. Walter Ltd., 26 Grafton Road, Worthing, Sussex, England, 1976), and *The Great Brain Robbery* (Chicago: Moody Press, 1976).

Wheeler, Gerald W., *The Two-Taled Dinosaur* (Nashville, Tenn.: Southern Publishing Assoc., 1975).

Wilder-Smith, A. E., *Man's Origin, Man's Destiny* (Marshall, Morgan and Scott Publications, Ltd., 116 Baker St., London W 1 M2BB, 1974; and Wheaton, Ill.: Harold Shaw Publishers, 1968), and *The Creation of Life* (Wheaton, Ill.: Harold Shaw Publishers, 1970).

Williams, Emmett L., co-authored by George Mulfinger, *Physical Science for Christian Schools* (Bob Jones University Press, Greenville, S.C. 29614), 1975, and "Entropy and the Solid State," in Walter E. Lammerts, ed., *Why Not Creation?* (Nutley, N.J.: Presbyterian and Reformed Publishing Co., 1970), pp. 67-69.

Wysong, R. L., *The Creation-Evolution Controversy* (Inquiry Press, P.O. Box 1766, East Lansing, Mich. 48823, 1976).

Zimmerman, Paul A., "Some Observations on Current Cosmological Theories," *Concordia Theological Monthly* 24:7 (July, 1953); editor, *Darwin, Evolution, and Creation* (St. Louis: Concordia Publishing House, 1959), and editor, *Creation, Evolution, and God's Word* (St. Louis: Concordia Publishing House, 1972).

B. Theologians and Philosophers

Berkhof, Louis, *Systematic Theology* (Grand Rapids: Wm. B.

Eerdmans Publishing Co., 1955), pp. 150-64.

Clough, Charles A., "Biblical Presuppositions and Historical Geology: A Case Study," *JCR* 1:1 (Summer, 1974), 35-48; *Laying the Foundation*, rev. ed. (Lubbock Bible Church [3202 - 34th St., Lubbock, Texas 79410], 1977).

Davis, John J., *Paradise to Prison: Studies in Genesis* (Grand Rapids: Baker Book House, 1975).

Fields, Weston, *Unformed and Unfilled: A Critique of the Gap Theory* (Nutley, N.J.: Presbyterian and Reformed Publishing Co., 1976).

Lang, Walter, ed., *A Challenge to Education* (Caldwell, Idaho: Bible-Science Assoc., 1974).

Leupold, H. C., *Exposition of Genesis* (Columbus, Ohio: The Wartburg Press, 1942).

Nelson, Byron, *After Its Kind*, rev. ed. (Minneapolis: Bethany Fellowship, Inc., 1967).

Rehwinkel, Alfred M., *The Flood* (St. Louis: Concordia Publishing House, 1951), and *The Wonders of Creation* (Minneapolis: Bethany Fellowship, Inc., 1974).

Rushdoony, Rousas J., *The Mythology of Science* (Nutley, N.J.: The Craig Press, 1967).

Scaer, David P., "The Problems of Inerrancy and Historicity in Connection With Genesis 1-3," *Concordia Theological Quarterly* 41:1 (January, 1977), 21-25.

Whitcomb, John C., *The Genesis Flood*, co-authored with Henry M. Morris (Nutley, N.J.: Presbyterian and Reformed Publishing Co., 1961), *The Origin of the Solar System* (Nutley, N.J.: Presbyterian and Reformed Publishing Co., 1963), *The Early Earth* (Grand Rapids: Baker Book House, 1972), *The World That Perished* (Grand Rapids: Baker Book House, 1973), and *Chart of the Period from the Creation to Abraham* (Winona Lake, Ind.: BMH Books, 1977).

Index of Subjects

(Indexes compiled by William Darr)

Aegean Sea, 149
Akkadians, 158
American Space Telescope, 31
Angular momentum: of earth, 38, 39
Antediluvian vapor canopy, 97
Apollo project, 9, 13, 14, 19, 30, 31, 33, 34, 36, 42, 86, 88, 90, 95, 98, 145
Arabia, 161
Assyrian eponym canon, 134, 135
Asteroids: belt, 138; interstellar, 97; particles, 48

Babylon, 159-161
Babylonian captivity, 147
"Bad theory": definition of, 63
Betelgeuse, 129
Biblical creationism, 64
Biosphere: problem explaining, 57
"Break-away theory": see Lunar origin: by fission
Bur-Sagale, 134

Caesarea of Palestine: sewer system, 140
California: educational controversy, 59
Cambrian period, 94
Canaan, 155, 160, 162
Canon: Assyrian eponym, 134, 135
Canon of Ptolemy, 134, 135
Carpathians, 113
Catastrophe, global: evidence of, 57
Catastrophes: Biblical perspective of, 97
Catastrophism: renaissance of, 58
Celestial mechanics: laws of, 35
Cenozoic era, 94
Cobrahead, 114
Copernican revolution, 14
Cosmic evolution: scientific obstacles to, 56

Cosmogony: definition of, 56; evolutionary, 17; Near Eastern mythological, 162; need for special revelation, 69
Cosmologies: Ancient Near Eastern, 65; definition of, 56; need for special revelation, 69
Craters: Alphonsus, 108, 114, 117, 118; Aristarchus, 106, 108, 111, 114, 116, 121, 122; Bullialdus, 120; Calippus, 110; Clavius, 86; Herodotus, 114, 116; Oceanus Procellarum, 86, 88; Peirce A, 120; Plato, 108, 112, 114, 116, 117
Creation: 13; Biblical doctrine of, 15; order of, 72
Creation/Curse/Flood: model of Genesis, 64
Creation Research Society, 23, 59
Creation Science Research Center, 59
Creation week, 77, 78
Creationism: scientific, 21

"Dark side" phenomena, 114
"Daughter" theory: see Lunar origin: by fission
Day-age theory, 74
"Dead planet," 122
Deluge, global: forces of, 64
Descartes highlands, 88
Doctrine: Biblical definition of, 14
"Double planet," 131
"Double-revelation" theory: 17; advocates of, 163; denial of supreme authority, 62; explanation of, 54, 55; harmonization attempts, 65; problem with earth's magnetic field, 61; scientific problem with, 57

"Earthshine," 122
Eclipse, solar: 132-136; earth moon
    compared to other moons, 133; in
    Assyrian history, 134; mentioned
    by Ptolemy, 135; significance for
    Biblical studies, 135
Egypt, 157
English Channel: tides, 140
Enlil, 158, 159
Evening: Biblical use of term, 77
Existentialism: neo-orthodox, 14
Ex nihilo, 73, 75n, 83, 162

"Fission" theory: see Lunar origin:
    by fission
Flood: God's promises about, 147
Fra Mauro, 107

Grammatical/historical interpretation:
    literary form in Genesis, 62, 63, 72;
    of Biblical account of creation, 55,
    72, 76

Harran, 159, 160
"Harvest moon," 141, 143
Hazor, 160, 161
Heraclides Promontory, 113
Hermeneutics: see Grammatical/historical
    interpretation
Historical/grammatical interpretation: see
    Grammatical/historical interpretation
Hittites, 159

Institute for Creation Research, 59
Interstellar gas clouds, 47

Jericho, 160
Jerusalem, 147, 150, 157
Judah, 147, 157
Jupiter, 43, 48, 111, 121, 130, 138
Jura mountain range, 113

Kantuzilis, 159n
Khonsu, 160
Koran, 62

Leibniz mountains, 113
Lick Observatory, 122
"Limited uniformitarianism," 147
Literal day interpretation: 76, 77;
    Biblical support for, 79; opposition
    to, 79, 81
London: water pumps of, 140
Lowell Observatory, 114
Luna 3, 31
Lunar: see Moon
Lunar albedo: definition of, 145
Lunar Alps, 112, 122

Lunar atmosphere, 85
Lunar dating, 91-103
Lunar gravitational field, 86
Lunar inclination, 141
Lunar mountain range, 85
Lunar origin: by capture, 36, 42-45;
    description of theory, 42; seen as
    doubtful, 45
Lunar origin: by condensation, 36,
    45-48
Lunar origin: by fission, 36, 37-42, 48,
    141; problems with, 38, 45
Lunar origin hypotheses: weakness of
    naturalistic, 50
Lunar origin theories: Apollo preference
    poll, 51; search for natural model,
    30
Lunar phases, 141
Lunar properties, 140-144
Lunar rock: breccia, 87-89; description
    of, 88
Lunar rock: crystalline, description
    of, 87, 88
Lunar rock: regolith, 87
Lunar rock analysis: compared to
    earth's, 89, 90; lack of carbon
    and free oxygen, 90
Lunar rocks, 31, 98
Lunar soil: 87; description of, 88
Lyellian/Neo-Darwinism: models of
    geology and paleontology, 57

Magnetic field: earth's: decrease in,
    59; implication for "young earth,"
    59, 62; measurement of, 59
"Man-in-the-moon," 88
Marduk, 160
Mares: 95; Crisium, 106, 119, 120;
    Fecunditatis, 119; Humorum, 119;
    Imbrium, 88, 112, 113, 119;
    Nectaris, 119; Nubium, 119, 120;
    Serenitatis, 119; Tranquillitatis,
    119; Vaporum, 119, 120
Mars, 30, 48, 97, 130, 138
Medo-Persian Army, 162
Mercury, 96
Mesopotamia, 158
Mesozoic era, 94
Metaphysics: need for special
    revelation, 69
Meteorites, 46
Milky Way: description of, 129
Mont Blanc, 112
Moon: see Lunar
Moon: Biblical references to, 53, 70, 71,
    147-151; Biblical revelation of, 70;
    brightness of, 132; condition at

second coming of Christ, 149-151; creation of, 67, 69, 73-83; dark side, 110; dependability of, 149; distance from earth, 30; dust, 94, 95; Hebrew concept of, 71; part of calendar, 148; present activity of, 105; shape of, 136-138; size of, 131, 136-138; theological significance, 153-155
Moon-Blink, 107
Moonquake epicenter: definition of, 120
Moonquakes, 31, 123
Moon worship: Ancient Near East, 157-162; Old Testament message against, 154-157
Morning: Biblical use of term, 77

Nabonidus, 158
Nanna, 158, 162
Nannar: see *Nanna*
NASA: program costs, 33
Natural law, 10
Naturalistic theory: 10; concerning origin of man, 36
Nebular theory: see Lunar origin: by condensation
Neo-Babylonian Empire, 160
Neo-Darwinists, 63
Neptune, 130, 133n

Ocean tides: 138-140; benefits and uses of, 140; causes of, 139
Orbital angular momentum: description of, 141
Orbital eccentricity, 44
Orbital synchronism, 145
Orion, 129

Paleozoic era, 94
Palestine, 161
Palomar, Mount: telescope, 30, 87
Phobos, 130
Photoelectric photometry, 120
Photosphere, 134
Pico, 112
Piton, 112
Pluto, 114, 133n
Pseudo-theologians: Christ's condemnation of, 68
Ptolemy: canon of, 134, 135

Radiometric dating: 93, 98, 99, 101, 102; Carbon-14, 93; daughter nuclei, 98, 99, 101; parent nuclei, 99, 101; Rubidium-Strontium, 96, 99
Rance River (France): electric generator, 140

"Red spots," 107, 114, 116
Retrograde orbit: 40, 48; examples of, 48
Revelation: direct, see Revelation: special
Revelation: general, in nature, 54
Revelation: indirect, see Revelation: general
Revelation: special, 54, 73; clarity of, 62; in Scripture, 54, 62
Rilles: description of, 86
"Rising of the sun," 67
"Roche's Limit": description of, 41

Sabbath: length of, 81, 82; reason for, 81
Saturn, 41, 43, 48, 113, 130
Scaling problem: description of, 47
Schröter's Valley, 114, 116
Scientific method: in historical sciences, 56; limitations of, 55
Scientist: definition of, 54
Scriptures: Christian, 14
Scriptures: Jewish, 14
Sea of Tranquillity, 31
Second law of thermodynamics, 63
"Secondary" planet, 131
Seventh Annual Lunar Science Conference, 50
Shamash, 158
Simanu: month of, 134
Sin: 158-160, 162
Sinai, 162n
Sirius, 129
Sister theory: see Lunar origin: by condensation
Solar system: creation of, 72
"Spin offs," 33
Sputnik, 29
Straight Wall, 120, 124
Sudden transformations: examples of, 75
Sumer, 158
Supernova, 101
"Survival of the fittest," 63

Taurus Mountains, 113
Tema, 161
Teneriffe Mountains, 112
Terminator: definition of, 112
Thebes, 160
Theologian: definition of, 54
Third Lunar Science Conference, 36
Tidal friction: 40, 44; energy loss from, 44
Tidal stress, 123
"Tracking phenomenon," 162

Transient lunar phenomena: area of
   greatest activity, 116-122; causes
   of, 121; correlation with charged
   particles, 125; evidence for, 107;
   historical background, 105-110;
   method of energizing, 123; method
   of observation, 109; method of
   reporting, 109; number of sightings,
   108; on "dark" side, 110; probable
   source, 127; reality of, 126; relation
   to volcanic activity, 123; serendipity
   of observations, 109; spectrograms
   of, 117, 118; terrestrial dust storms,
   126; view by scientific community,
   106
Triton, 130
Turkey, 159

Ultimate contradiction, 15
Ultimate origins, 67
Uniformitarianism: 39; level of
   thinking, 80
Units of time: in Genesis, 78
Ur, 158
Uranus, 48, 110, 111, 121, 130

Venus, 30, 97, 132, 145
Viking, 30
Volcanoes: description of, 111;
   Herschel's hypothesis, 121; on
   moon, 111-113

Yarikh, 160

Ziggurat, 158, 159
Zion, 150

Index of Names

Abell, George, 44n, 108n, 131n
Abraham, 159, 160, 162
Adam, 148
Adams, John, 130
Ager, Derek A., 58n
Aldrin, Edwin E., 89
Alter, Dinsmore, 90n, 117, 118n
Anders, William, 29
Armstrong, Harold L., 64n, 76n
Armstrong, Neil A., 31, 89
Ashbrook, J., 121n
Ashurbanipal, 162n
Asimov, Isaac, 95n

Bahnsen, Greg L., 56n
Baldwin, Ralph B., 38, 39, 44, 50, 107, 117n
Barnes, Thomas G., 59, 61
Barr, 111, 114
Beek, Martin A., 158n, 159n
Beer, Wilhelm, 107, 117
Belshazzar, 158, 161, 162
Berkhof, Louis, 56n, 82n
Bethell, Thomas, 56n
Bianchini, Francesco, 116
Bickerman, Elias J., 148n
Bildad, 15
Bliss, Richard B., 65
Borman, Frank, 29
Bouw, Gerardus D., 21n
Boyer, James L., 148n
Boyle, Robert, 16n, 53
Brant, J. C., 47n
Briggs, Charles A., 77n, 82n
Brown, Francis, 77n, 82n
Budine, 114
Bullinger, Ethelbert W., 72n
Burgess, Eric, 130n
Burley, J., 107n
Butterfield, Lyman H., 130n

Cameron, Alastair, G. W., 39n, 94n, 141n
Capen, 120
Caroché, 113
Cassini, Jacques, 119
Catacosinos, Paul A., 98n
Cattermole, Peter, 107, 113
Chaffee, Roger, 33
Chevalier, 111
Cisne, J. L., 38n
Clark, Gordon H., 57n
Clark, Robert, E.D., 16n
Cleminshaw, Clarence H., 90n
Cloud, Preston, 21n
Clough, Charles A., 57n, 71n
Collins, Michael, 89
Compton, Arthur H., 129
Copernicus, Nicolaus, 66
Corliss, William R., 125, 126n
Cortright, Edgar M., 42n
Cyrus the Great, 162

Dachille, Frank, 58n, 97n
Darwin, Charles, 37
Darwin, George H., 37
Davis, John J., 57n, 72n
Deitz, Robert S., 58n
de Magellan, Jean H., 110
de Vaucouleurs, Gerard H., 108n, 136n
de Wit, J. J. Duyvene, 57n
DeYoung, Donald B., 22, 23, 28, 99n
Dickman, Robert L., 47n
Docherty, 113
Dostol, K, P., 99n
Driver, Godfrey R., 158n
Driver, Samuel R., 77n, 82n
Dubberstein, Waldo H., 134n, 148n
Dudley, H. C., 98n
du Noüy, Lecomte, 56n

Eberhart, J., 130n
Ecklemann, Herman J., Jr., 47n, 77n,
    81n, 82n
Edwards, Frank, 106, 107
Ehmann, William D., 36, 91n
Eldredge, Niles, 58n
Emmett, 119
Ethan the Ezrahite, 148
Eysenhard, 119

Fallows, Fearon, 111
Faraday, Michael, 16n, 125
Farrell, 114
Fields, Weston, 57n, 72n, 73n, 82n
Fimmel, Richard O., 130n
Finegan, Jack, 158n
Fink, U., 130n
Flammarion, Camille, 30, 117
Fort, Charles, 106
Fotheringham, John K., 148n
Franks, 113
French, Berian M., 51, 94n
Friedman, Herbert, 134n

Galileo, 30, 66, 85
Garfinkle, S. B., 102n
Gast, Paul, 91
Gautier, T. N., 130n
Gentry, Robert V., 98n
Gerling, Christian L., 112
Gerstenkorn, H., 41n
Gingerich, Owen J., 93, 94n
Goetze, Albrecht, 159n
Gold, Thomas, 45, 95n
Goldreich, Peter, 38n, 40n, 42, 43n,
    91, 93n, 140n
Gould, Stephen Jay, 58n
Grassé, Pierre P., 58n
Greenacre, 111, 114
Grene, Marjorie, 56n
Griffiths, J. Gwyn, 160n
Grissom, Virgil, 33
Gruithuisen, Franz P., 111, 116, 120
Guinness, Os, 153n
Gunst, R. H., 61n

Haggerty, James J., 33n
Hammond, Allen L., 35n, 39, 40n, 41,
    43
Hammurabi, 158n, 159
Harris, 116
Harris, A. W., 41n
Harris, R. Laird, 66n, 67n
Hart, 113
Hart, R., 133n
Hartmann, William K., 95n, 97
Hauksbee, Francis, 126
Heribert-Nilsson, N., 58n

Herschel, William, 110, 111, 112, 114,
    121, 122
Hezekiah, 156
Himmelfarb, Gertrude, 56n
Hodgson, 112, 113
Holden, Edward, 112, 122
Hooykaas, Reijer, 15, 16n, 66n, 105,
    154n
Howe, George F., 58n, 64n
Hummel, Charles E., 66n

Ingall, 119
Isaiah, 156, 157

Jackson, John G., 119
Jacobs, Kenneth C., 138n
Jaki, Stanley L., 56n
Jeffries, Harold, 38, 39
Jeremiah, 156, 157
Job, 155
Joel, 149
John, 149, 150
Josephus, 140n
Joshua, 160
Josiah, 157
Judson, Sheldon, 124

Kaiser, Walter G., Jr., 62, 63n
Kaplan, Martin M., 58n
Kaula, W. M., 41n
Kelly, Allan O., 58n, 97n
Kelvin, Lord, 16n
Kent, Homer A., Jr., 81n
Kepler, Johannes, 16n, 131n
Kerkut, G. A., 35, 58n
Kidner, Derek, 156n
Kloosterman, Johan B., 58n
Klotz, John W., 55n, 66n
Kopal, Zdenek, 136, 137n
Kozyrev, Nikolay A., 107, 117, 118,
    122, 124
Kuiper, Gerard P., 46

Lamb, Horace, 61
Lambert, Johann H., 113
Langdon, Stephen H., 148n
Langford, Jerome J., 66n
Larson, H. P., 130n
Leet, Florence J., 124n
Leet, Lewis Don, 124
Le Gallic, Y., 102n
Leitch, Addison H., 79n
Leupold, Herbert C., 57n, 79n, 156n
Levi-Strauss, Claude, 58n
Lewy, Julius, 162n
Ley, Willy, 107, 111n
Lindsell, Harold, 57n
Longacre, Robert E., 56n

Lovell, James, 29
Luther, Martin, 66
Lyot, Bernard, 134
Lyttleton, Raymond A., 95n

Maatman, Russell W., 74, 77n, 81n
Macbeth, Norman, 56n, 64n
MacDonald, Gordon J. F., 39, 40, 41n, 47n
Mädler, Johann H., 107, 117
Mann, Wilfrid Basil, 102n
Marsden, B. G., 39n, 94n, 141n
Mathisen, James, 160n
Matthiae, Paolo, 62n
Maxwell, James Clark, 16n
McClain, Alva J., 150n
McCorkle, 113
McDonald, K. L., 61n
McKinzie, R., 44n
Medillo, M., 133n
Meek, Theophile J., 158n
Menzel, Donald H., 108n, 136n
Middlehurst, Barbara M., 107n, 113n, 122, 123
Miles, John C., 158n
Mills, A. A., 125, 126
Mitroff, Ian I., 50n, 51n
Moore, 112, 113
Moore, Patrick A., 107, 120, 122, 123
Moorehead, Paul S., 58n
Morris, Henry M., 22, 57n, 58, 61n, 64n, 65, 67n, 72n, 82n, 83n, 96n, 103n, 147n
Morris, Leon, 78
Morse, Samuel F. B., 16n
Moses, 155, 157, 162
Mulfinger, George, Jr., 21n, 23, 47n, 64n
Munk, Walter H., 38n, 39, 40n
Mutch, Thomas, 95n, 97n

Nabonidus, 158, 160, 161n, 162n
Nagel, M., 99n
Nebuchadnezzar, 157, 160
Nelson, Byron, 57n
Neugebauer, Otto, 135
Newell, Norman D., 21n, 58n
Newman, Robert C., 47n, 77n, 81n, 82n
Newton, Isaac, 102, 131n
Newton, Robert Russel, 102n
Nordmann, J. C., 38n
North, Gary, 56n, 69n, 71n

O'Hair, Madalyn Murray, 18
Olbers, Heinrich, W. M., 120
Oppenheim, A. Leo, 160n, 161n
Ordway, Frederick I., 33n

Pabst, D., 99n

Pacer, Richard A., 36, 91n
Parker, Richard A., 134n, 148n
Pasachoff, Jay H., 95n
Paul, 159
Payne, J. Barton, 62n
Penny, David, 64n
Pettinato, Giovanni, 62n
Pfeiffer, Charles F., 158n
Phillips, John Gardner, 90n
Piazzi, Giuseppe, 111
Pickering, William, 112
Pliny, 126
Popper, Karl, 56n, 63, 64, 85
Poythress, Vern S., 57n, 71n
Priest, Joseph, 140n
Pritchard, James B., 158n, 159n, 160n
Ptolemy, Claudius, 102n, 134, 135

Randall, John L., 76n
Rankin, 111
Rawstron, 112
Read, John G., 98n
Redekopp, Larry G., 11
Rehwinkel, Alfred W., 57n
Reymond, Robert L., 57n
Roche, Edouard, 41n
Rudge, 125
Rushdoony, Rousas J., 57n
Ruzic, Neil P., 33n

St. Peter, Robert L., 64n
Sanduleak, 120
Schaeffer, Francis A., 57n, 153n
Schindewolf, P. H., 58n
Schmidt, Johann, 112
Schneller, 112
Schröter, Johann H., 111, 112, 116, 119, 120
Sennacherib, 162n
Serviss, Garrett P., 121
Shoemaker, E. M., 95n
Short, 116
Short, Nicholas, 108, 123n
Sinsharishkun, 162n
Slusher, Harold, 96n
Smith, Elske P., 138n
Solomon, 148
Spilsbury, Richard, 56n
Spykman, Gordon J., 57n, 65n
Standen, Anthony, 56n
Stansfield, William D., 56n
Stephens, 116
Stephens, Ferris J., 158n
Stock, 120
Stretton, 111
Swindell, William, 130n
Szebehely, V., 44n

Tan, Paul Lee, 72n
Taylor, S. Ross, 36
Teller, Woolsey, 18, 136n, 145, 146n
Thiele, Edwin R., 134, 135
Thomas, 120
Thompson, William R., 58n
Thomsen, Dietrick E., 103n
Tolansky, S., 89n
Tombaugh, Clyde, 114
Treash, Robert, 50n
Trippett, Frank, 13n, 56n
Turcotte, D. L., 38n

Urey, Harold, 50

Valier, 112
Van Til, Cornelius, 69n
Velikovsky, Immanuel, 101n
Verduin, Leonard, 75n
von Weizsäcker, C. F., 46
Vos, Howard F., 158n

Waltke, Bruce, 73n
Watkins, Norman D., 102n
Weaver, Kenneth F., 87n
Webb, Thomas W., 117, 121n, 119n
Wells, John W., 94
Welther, B., 107n
Westfall, Richard S., 102n
Whipple, Fred L., 108n, 125, 136n
Whitcomb, John C., 16n, 22, 23, 28, 55n,
    57n, 58, 64n, 72n, 73n, 81n, 83n,
    96n, 135n, 147n, 155n, 160n, 162n
White, Andrew, 16n
White, Edwin, 33
Wilder-Smith, Arthur E., 64n
Wilkins, H. P., 120
Wilkins, W., 111
Williams, Emmett L., 64n
Williams, W. O., 112
Wonderly, Daniel, 103n
Wood, John A., 91n
Woodward, 112
Worrad, Lewis H., Jr., 55n
Wright, Robert C., 99n
Wyatt, Stanley P., 86n

Yadin, Yigael, 161n
Young, Davis A., 59n, 80n, 81n
Young, Edward J., 57n, 62

Index of Scripture

Genesis
1—77, 78, 79, 80,
    81, 153, 154, 162
1/2—74
1/11—14, 63
1:1—29, 71n, 72n,
    74
1:1-2—71
1:1-3—73n
1:2—154n
1:3—73, 74
1:5—77
1:11—74
1:12—74
1:14—77, 78, 150
1:16—34, 53, 67, 70,
    72n, 73n, 74, 77,
    132
1:18—77
1:21—72n, 74
1:25—72n, 74
1:26—72n
1:27—72n, 74
1:28—16, 54
2:3—82
2:4—72n, 77
2:7—74
2:19—74
2:22—74
2:23—82n
3:14-24—97
3:17—82
4:23-24—63
5—83n
7:3—81n
7:11—97
8:22—147, 150
11—83n
12/50—63

12:1-3—147
18:32—82n
32:24—82n

Exodus
20:9-11—82
20:11—72n, 78
31:17—78
35:3—82

Numbers
7:12-78—76
17:8—74

Deuteronomy
4:19—156
7:7-8—147
10:10—149
11:11—90n
17:3—156
28:11—149

Joshua
10:12-13—150

Judges
8:21—156
8:26—156

I Kings
4:29-34—148
4:31—148
7:23—67n

II Kings
4:18-20—156
21:3-5—157
23:5—157

II Chronicles
4:2—67n

Esther
Book of—74

Job
12:15—90n
15:15—155
25:5-6—155
31:26—70
31:26-28—155
38:4—68

Psalms
8:3-4—71
10:6—66n
16:8—66n
19:1—70
19:1-2—68
19:4-6—67
30:6—66n
33:6—73
33:9—73
36:9—54
55:22—66n
69:2—66n
69:9—66n
72:5—148
78—63
89:37—148
90:4—80
90:6—77, 78
93:1—66n
96:10—66n
104:19-20—71
104:24—71
104:31—71
105—63

106—63
119:105—69
121:3—66n
121:4-7—156
135:7—90n
136—63
136:8—71
136:9—71
148:3—73n
148:5—73n

Proverbs
1:7—69
25:2—16

Ecclesiastes
1:7—90n

Song of Solomon
6:10—70

Isaiah
3:18—156
24:23—150
30:26—150
40:25—80
40:25-26—151
40:26—73n
40:26-28—68
40:28—81
40:28-29—80
45:22—151
49:10—156
51:13—17
55:9—80
55:10—90n
60:19-20—150

Jeremiah
7:16—157
8:1-2—157
9:1—157
10:2—156
13:17—157
31:35-36—147, 148
31:37—34
32:17—80
33:20—147
33:25—147

Ezekiel
32:7—149

Daniel
4:17—74
5—162
8:14—77
8:26—77

Joel
2:10-11—149
2:30-31—149
3:15—149

Amos
5:36—156

Jonah
4:6-10—74
4:8—156

Habakkuk
3:11—150

Zechariah
14:7—76

Matthew
1:8—83n
4:24—156
5:18—15
11:28-30—81
17:15—156
22:29—68
24:29—149

Mark
2:27—81
4:39-41—76
8:25—75
13:24-25—149

Luke
21:25—149
23:44-45—150

John
1:3—75
1:10—75
1:14—75
1:19/2:1—78
2:11—75
5:17—81
8:32—69
10:35—15
14:2—75n
20:31—75

Acts
2:19-20—149

Romans
1:18-20—68
1:18-23—67n, 146
3:11—67n
4:17—73n
5:12—97
8:20-22—97

I Corinthians
1:19-29—67n
2:9-10—69
2:14—67
15:41—70, 129

II Corinthians
4:4-6—74
5:17—74
10:5—69

Galatians
3:16-18—147

Colossians
1:16—75
2:3—23, 69
2:16—148

II Timothy
3:16—14, 15

Hebrews
1:2—75
4—81
4:3-5—81n
4:3-11—81
11:1-6—67n
11:3—16, 64, 73, 73n

II Peter
1:21—15
3:3-5—67n
3:8—79, 80
3:16—81

I John
4:3—74

II John
7—74

Revelation
6:12—150
7:1—67
7:2—67
8:12—150
20:11—150
21:1-2—150
21:23—150
21:25—150